THE
SCIENCE
digest
BOOK OF
HALLEY'S
COMET

THE SCIENCE digest BOOK OF HALLEY'S COMET

JOHN TULLIUS

AVON
PUBLISHERS OF BARD, CAMELOT, DISCUS AND FLARE BOOKS

THE SCIENCE DIGEST BOOK OF HALLEY'S COMET is
an original publication of Avon Books. This work has never
before appeared in book form.

AVON BOOKS
A division of
The Hearst Corporation
1790 Broadway
New York, New York 10019

Copyright © 1985 by John Tullius
Published by arrangement with the author
Library of Congress Catalog Card Number: 84-45579
ISBN: 0-380-89527-7

First Avon Printing, March, 1985

AVON TRADEMARK REG. U.S. PAT. OFF. AND IN
OTHER COUNTRIES, MARCA REGISTRADA, HECHO EN
U.S.A.

Printed in the U.S.A.

DON   10 9 8 7 6 5 4 3 2 1

For Lucille Aurora Des Jardins Tullius, who conceived of it all

ACKNOWLEDGMENTS

Thanks go to Alan S. Wood of the Jet Propulsion Laboratory, Neil Passey of the Hansen Planetarium, Helen Miller of the Lick Observatory, and Ray Newburn of the International Halley Watch for their generous assistance.

Special thanks are due to Mary Ann Rovai, H.B. Laski, and Tony Fish for their help in organizing the final manuscript.

Finally, thanks to Shannon, whose love made it easy.

TABLE OF CONTENTS

INTRODUCTION   11

PART ONE
COMET FEVER   15

CHAPTER 1:   THE MYTH, LEGEND, FOLKLORE,   17
                AND FEAR OF COMETS

CHAPTER 2:   COMET FEVER AND SCIENTISTS   25

CHAPTER 3:   WHAT IF A COMET STRUCK THE   28
                EARTH? AND OTHER PARANOIAS

PART TWO
THE STORY OF EDMUND HALLEY AND HIS COMET   31

CHAPTER 1:   THE UNIVERSE BEFORE NEWTON   33

CHAPTER 2:   THE WORLD ACCORDING   37
                TO NEWTON

CHAPTER 3:   A STORM OF JEALOUSY   47

CHAPTER 4:   COMETS BEFORE HALLEY   52

CHAPTER 5:   HALLEY'S WORK ON COMETS   56

CHAPTER 6:   *THE LEGACY OF*   69
SIR EDMUND HALLEY

PART THREE

ALL ABOUT COMETS   73

CHAPTER 1:   *COMET FAMILIES*   75

CHAPTER 2:   *HEADS AND TAILS: THE*   78
STRUCTURE OF A COMET

CHAPTER 3:   *WHAT A COMET IS MADE OF*   88

CHAPTER 4:   *WHERE DO COMETS COME FROM?*   93

CHAPTER 5:   *THE LIGHT OF A COMET*   98

CHAPTER 6:   *LOST COMETS, DEAD COMETS,*   101
AND METEORS

PART FOUR

THE RETURN OF HALLEY'S COMET   109

CHAPTER 1:   *HALLEY'S COMET THROUGHOUT*   111
HISTORY

CHAPTER 2:   *SPACE MISSIONS TO HALLEY'S*   119
COMET

CHAPTER 3:   *HALLEY'S COMET EVENTS,*   123
1985-86

FOOTNOTES   129

GLOSSARY   131

SUGGESTED READING   133

INTRODUCTION

There are few sights as impressive or as memorable as a fully developed comet with its tail stretched out across the sky. But, because a really brilliant comet appears so infrequently (on average, once every three or four generations), human beings tend to forget how awesome and powerful a comet can sometimes appear. The Great Comet of 1811 had a head (the bright ball of light from which the tail extends) that was 1.25-million miles in diameter—which is twice the size of the sun. Its tail was 15-million miles wide and 100-million miles long. It was visible to the unaided eye for nearly a year.

Although Lexell's Comet of 1770 wasn't actually as large as the 1811 comet, it seemed much bigger because it passed much closer to the earth. Its head looked five times larger than the full moon, and its tail stretched westward from the horizon past the mid-sky. But both the Great Comet of 1811 and Lexell's Comet were dwarfed by the Comet of 1843, the largest ever recorded. The tail of this one stretched 200-million miles, which is more than twice the distance between the earth and the sun, and at its most brilliant it filled nearly one third of the night sky.

Not only are comets of awesome size, length, and brilliance but also they come in a selection of colors. Although most comets are white, they have been recorded as gold, red, blue, and even green, and when they are close to the horizon they will take on the colors of the sunset. This leads to breathtaking displays of pastel comets with tails every color of the rainbow. And if that's not enough, comets often have more than one tail. De Chéseaux's Comet of 1744 was probably the most spectacular ever observed. It had at least six bright broad tails that spread out "like a great astronomical fan" against the sky.

The enthusiasm for the coming of Halley's Comet 1985–86 could be unprecedented. When Comet Kohoutek, a rather unimpres-

11

sive, barely visible comet, flew by in 1973, the Hayden Planetarium in New York received more than 50,000 phone calls and letters in a three-week period.

Halley's Comet will be many times brighter and more spectacular than Kohoutek's. It will also have history on its side. Halley's is not just some anonymous cosmic event; it is intricately woven with human tradition.

Halley's Comet is etched in human annals from the death of Julius Caesar to the Battle of Hastings. It has acted like a sort of cosmic clock that tolls every 76 years or so. It's something a man tells his children and grandchildren about. Your grandparents, in fact, probably saw Halley's in 1910 and, like a prophecy fulfilled, you will see it. Most likely your grandchildren will be around for its appearance midway through the 21st century.

Halley's Comet is, indeed, for all to wonder at—not just the scientists and astronomers who will so laboriously study in detail its coming and going. This book is for the layman who wants to know something about this spectacle that we are about to witness and enjoy.

THE
SCIENCE
digest
BOOK OF HALLEY'S COMET

PART ONE
COMET FEVER

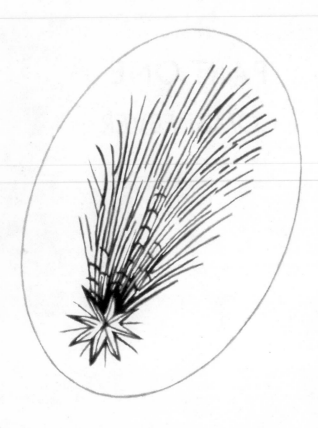

This drawing of Halley's Comet, reproduced from the Chronicles of the city of Nuremburg in A.D. 684, is the oldest known picture of a comet.

CHAPTER 1

THE MYTH, LEGEND, FOLKLORE, AND FEAR OF COMETS

Months before Halley's Comet becomes visible in the night sky in the winter of 1985–86, the world will be suffering an epidemic known as comet fever.

Seeing a comet has always done something strange to the human psyche. Put yourself, for instance, in the place of a man of 10,000 years ago who, wakened by his wailing wife, stumbles out of his cave one night to behold a bright ball of light with a fluorescent tail stretching halfway across the heavens. It hangs there for several months while it creeps eastward across the sky, then disappears. But after a few weeks it *reappears,* with its tail pointing in the other direction, and proceeds to march back across the heavens like a funeral procession returning from a gravesite.

What emotion would you feel standing there in your bearskin without 500 years of science to reassure you? Awe? Inspiration? Amazement? Perhaps. But more likely you'd feel fear and dread and anxiety. And from that day forward you would probably blame every calamity that befell you, from a broken toe to the loss of a tribal fracas, on this ominous visitor.

Human beings have always looked to the sky for answers to the vagaries of life and predictions of the future. Because no heavenly events are more spectacular than a spectacular comet, the legends, superstitions, folktales, and fear surrounding comets have grown to spectacular proportions. Comets were believed to change history, kill off kings, and decide battles, and have been linked with every cataclysmic event from the Great Flood to the End of the World.

17

Nearly two thousand years ago the Roman scholar Pliny the Elder commented:

> We have, in the war between Caesar and Pompey, an example of the terrible effects which follow the apparition of a comet. Toward the commencement of this war, the darkest nights were made light, according to Lucan, by unknown stars, the heavens appeared on fire, burning torches traversed in all directions the depth of space; the Comet, that fearful star, which overthrows the powers of the Earth, showed its terrible locks.

This may seem a bit hysterical for a man of Pliny's reputed good sense, but his reaction was considered mild.

The prize-winner for hysterical comet reporting goes to Ambroise Paré, a famous French surgeon of the 16th century who was generally a sane and eminently sensible man (he became personal physician to four kings of France and developed several medical techniques still in use today). The Comet of 1528, however, drove him to this statement:

> This comet was so horrible and so frightful and it produced such great terror in the vulgar that some died of fear and others fell sick. It appeared to be of excessive length and was the colour of blood. At the summit of it was seen the figure of a bent arm, holding in its hand a great sword, as if about to strike. On both sides of the rays of this comet were seen a great number of axes, knives and blood coloured swords among which were a great number of hideous human faces with beards and bristling hair.

The blame for all this bad news was usually placed on—who else—the Devil. The legendary sulfurous odor of Hell was said by many to be present whenever a comet appeared. And, of course, wherever that smell was, the Devil, always a bad egg, was sure to be close at hand. In fact, the book *Cometomania* published in 1684 used the testimony of a group of monks who swore they could always smell a comet as proof that astronomers who suggested that comets inhabited space "way beyond the moon" were all dead wrong. This idea, of course, conveniently supported the long-held theory formulated by Aristotle around 350 B.C. that comets were nothing more than "hot, dry exhalations" rising up from the earth and sparked into flame by the frictions of the sky's movement. If you could smell the Devil's work, it was, of course, nonsense to suppose that comets dwelt beyond the regions of the moon. It also supported the medieval view of the Catholic Church, whose doctrine concerning the universe rested on Aristotelian physics.

The belief that comets were devil-sent probably reached its greatest absurdity, though, when Halley's Comet appeared in 1456. At that time the Turkish army, led by Mohammed II, a fearsome warrior, had for several days been besieging Belgrade (protected by a Christian army) when the comet suddenly came into view in the morning sky. The Christians, of course, were struck numb by impending doom, for they believed these pagan Turks were in league with the Devil. Luckily for Belgrade, the Turks had their own set of comet superstitions and they, too, were paralyzed by fear. Mean-

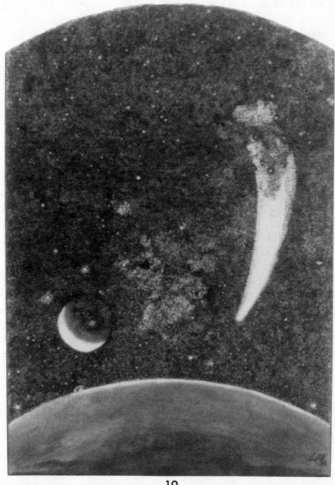

while, Pope Calixtus III ordered public prayers to be offered up for deliverance from the foes of Christianity and then, he excommunicated the comet! The Turks were almost immediately routed, which probably has nothing to do with the fact that their weakened forces had run out of provisions nearly a week before.

Comets have always suffered from bad press. And no wonder! It seemed that every time a great comet appeared, some great person—particularly a king or prince—was instantly dispatched. As Shakespeare put it: "When beggars die, there are no comets seen; The heavens themselves blaze forth the death of princes." (*Julius Caesar,* Act II, Scene 2.)

The author of the booklet, *Cometomania,* written in 1684, explained this well-known vulnerability of kings to comets in this way:

> Comets distemper and inflame the air and exhaust the juices of the Earth. . . . As the inevitable effects of both, we must expect sickness, diseases, mortality, and more especially the sudden death of many Great Ones, because these are sooner and more easily hurt than others, for their delicate feeding, and luxurious course of life, and sometimes their great cares and watching which weaken and infeeble their bodies, render them more obnoxious than the vulgar sort of people.

The comet of A.D. 453 supposedly presaged the death of Attila the Hun. The comet of 455 was supposed to have triggered Emperor Valentinian's death. In fact, every bright comet that appeared during Roman times, the Dark Ages, the early Medieval period, the high Middle Ages, and even the Renaissance was supposed to have announced the death of some great person.

Naturally, there were exceptions. When Charlemagne, the emperor of Western Europe, died in 814 there was no comet visible. This fact, however, did not dissuade the chroniclers of the period from recording one. Before that, when the bright comet of 44 B.C. appeared several months *after* the murder of Julius Caesar, the theory that comets *predicted* the death of kings was set conveniently aside and the comet was recorded as the victim's soul ascending into heaven.

Vespasian, a levelheaded, just, and benevolent Roman emperor, refused to be bullied by all this superstitious balderdash. When the comet of A.D. 79 rolled around, he boldly announced: "This hairy star does not concern me; it menaces rather the King of the Parthians, for he is hairy while I am bald." Vespasian died, nevertheless, within three months.

Nero, the infamous Roman Emperor, wasn't fiddling around when a bright comet appeared in A.D. 60. The historian Tacitus wrote: "As if Nero were already dethroned, men began to ask who might be his successor." The astrologer Balbillus advised Nero that the only way to "deflect the wrath of the heavens" was to kill off important persons and anyone who could succeed to his throne, thereby fulfilling the omen and eliminating competition for the throne all in one whack. As the historian Suetonius relates, Nero was quick to take Balbillus's advice: "Nero resolved on a wholesale massacre of the nobility. . . . All children of the condemned men were banished from Rome, and then starved to death or poisoned."

Nero did indeed survive the comet, and even the return of a Halley's Comet in A.D. 66. He survived long enough, in fact, to commit suicide two years later at the age of 32.

Another adherent to this theory of "do unto others before the comet can do you in" had been the Roman general Marius, who by comparison makes Nero look like an overachieving boy scout. When the comet of 87 B.C. appeared, Marius tortured, then crucified, then beheaded, then flayed every Roman aristocrat and any of their unfortunate sons who had ever opposed him as Consul. Marius died anyway, later that year.

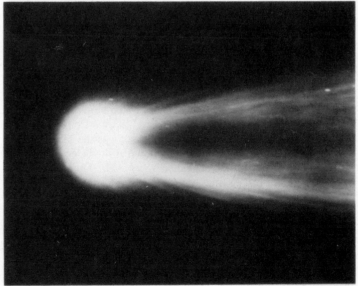

Close-up view of Halley's Comet in 1910 looks remarkably like the depiction on the Bayeux Tapestry. *Mt. Wilson Observatory.*

War has also provided suitable material for comet fever. In fact, one of western civilization's most celebrated battles coincided with perhaps the most celebrated return of Halley's Comet: When it appeared in April of 1066 it was viewed as an omen, and it was ominous indeed for King Harold of England. Harold was killed that year at Hastings, fighting the invading Normans. It was William the Conqueror's wife, back home in Normandy, who created the famous Bayeux tapestry, in which the decisive influence of the comet's power is depicted. The tapestry shows crowds of onlookers whose eyes are literally popping out at the sight of the comet while King Harold sits in shocked dismay on his throne.

Napoleon Bonaparte was a great believer in the portentous nature of comets. The comet that appeared in 1769, the year he was born, was thought to be his protecting genie. And Napoleon himself announced that he believed the Great Comet of 1811 was an omen for the eventual success of his great campaign into Russia. So much for omens.

Naturally, the greatest human calamities, such as the Great Plague, had to be connected with a comet. In his *Journal of the Plague Year,* Daniel Defoe wrote:

> A blazing star or comet appeared for several months before the Plague. . . . It passed directly over London so that it was plain it imported something peculiar to the city alone . . . it was of a faint, dull, languid colour, [and] its motion was very heavy and it accordingly foretold a heavy judgement, severe, terrible and frightful as was the Plague.

The paranoia surrounding comets is not limited to ancient times and the Dark Ages. At the last appearance of Halley's Comet, in 1910, there were many strange reactions. One such was the sacrificial rite of an occult band in rural Oklahoma, known as the Select Followers, who saw the comet as the vengeance of Jehovah. The local sheriffs arrived one night just in time to prevent these demented souls from sacrificing a virgin.

Most of the commotion surrounding Halley's Comet in 1910, however, centered around the fact that the earth was scheduled to pass through the tail of the comet: Research had revealed the presence of cyanogen, the lethal ingredient in cyanide, in the tail. Also, a French writer had suggested that hydrogen in the tail could react with the earth's atmosphere, "overwhelming our planet in a gigantic explosion."

The panic and debate that ensued was massive, and people set about watching the comet in homemade gas masks. Itinerant patent-

Halley's Comet on May 12, 1910, with a 30-degree tail and on May 15 with a 40-degree tail. Earth passed through the tail of the comet on its 1910 passage. *Mt. Wilson Observatory.*

medicine vendors, like "T.B." Jones, made their fortune selling "comet pills," which were supposed to protect you against every known comet-connected malady including—you guessed it—tuberculosis.

Some folks sealed their windows against cyanogen poisoning, while others kept watch as the earth plunged into the comet's tail. As James Thurber, then 16 years old, explained it:

Nothing happened, except that I was left with a curious twitching of my left ear after sundown and a tendency to break into a dog-trot at the striking of a match or the flashing of a lantern.

In 1973, scientists predicted that the approaching Comet Kohoutek would be as bright as any comet of the last few centuries and clearly visible in daylight, which would make it spectacular. That was all it took to set comet fever burning again. In the hysterical

1970s when gurus, prophets, and holy men were a dime a dozen, the predictions of doom surrounding Kohoutek were epidemic. In a booklet entitled *The Christmas Monster,* Moses David, of the organization called the Children of God, went so far as to say the comet would bring "the end of Fascist America and its new racist Emperor, Richard Nixon." He predicted the comet would grow seven times the size of the moon and bring about the destruction of the world.

In fact, Comet Kohoutek proved to be a major disappointment for scientists and gurus alike. With a small tail barely seen in the night sky, it remained visible to observation with binoculars for less than a month and then was gone—not to return for another 75,000 years.

The world did not end.

Comet Bennett of 1970, one of the most brilliant comets of this century, was considered by many Palestinian peasants to be a secret weapon of the Israeli army. *Lick Observatory photograph.*

CHAPTER 2

COMET FEVER AND SCIENTISTS

The scientific community has not been immune to cometomania, either. On the contrary, at times they've raised it to new levels of absurdity. Merely searching for comets has brought out an irrationality ordinarily reserved to the ignorant and the superstitious. Some astronomers have become so obsessed with the idea of discovering comets that all else in their lives becomes secondary.

Charles Messier (1730–1817) is probably the best example. Considered the greatest of comet hunters, he was obsessed with finding them. Unfortunately, around the time of the return of Halley's Comet in 1763, his wife had the misfortune of coming down with a fatal illness. While Messier was busy nursing his wife, his arch-rival, Montaigne of Limoges, found a most important comet. At his wife's funeral, when someone offered condolences, Messier, beside himself with grief, could only utter: "Yes, it is too bad. He has robbed me of my thirteenth comet!" Then, realizing what he'd said, he quickly added, "Ah, poor woman!"

Not long after the funeral, Messier was walking in his garden and constantly gazing at the sky on the off-chance of spotting a comet. Instead he fell headlong into a well, was nearly killed, and spent the next several months in the hospital.

Throughout history there have been scholars who cited comets as the cause of all types of unexplained happenings, from a particularly excellent wine harvest ("Comet Wine of 1811" was advertised for nearly a century in the catalogs of London wine merchants) to the theory (put forth in the 1970s by Sir Fred Hoyle and Chandra Wickramsinghe, both highly respected astrophysicists) that life on the planet Earth began with comets and that many diseases, such as tuberculosis, bubonic plague, and smallpox, developed from viruses and germs scattered here by passing comets.

In the 18th century, Maupertuis, the French mathematician, thought that comets "were peopled by a certain race of men" and that their tails contained "a dazzling train of jewels." If a comet collided with the earth, therefore, he thought "Earth would enjoy rare treasures which a body coming so far would bring to it. We should be much surprised to find that remains of these bodies we despise are formed of gold or diamonds."

Even the greatest cometologist of them all, Sir Edmund Halley himself, went a bit daft in theorizing about comets. For example, he was convinced that the Great Flood was due to "the casual shock of a comet." The shock altered the earth's axis of rotation, creating a rocking motion that made the oceans of the world slosh their waters onto the land in huge waves. He thought the Caspian Sea was probably the comet's crater. This idea of Halley's may seem a bit far-fetched for the man who cleared up most of the mystery and misconceptions surrounding comets. We should remember, however, that there was intense pressure on scientists in those days to align their theories with the dogma of the Church.

Many other theories over the centuries have linked comets with Biblical events. It was Seneca, in fact, in the first century who first said the Great Flood had been caused by a "wandering planet"—a comet. And in Halley's time there was a scholarly stampede to prove that the comet named for him had been the "star" of Bethlehem.

No one has been as thorough in his theoretical folly as Dr. William Whiston, an English scholar and clergyman, contemporary of Newton and Halley, who succeeded Newton as Lucasian Professor of Mathematics at Cambridge in 1710. In 1703, Whiston published a book entitled *A New Theory of the Earth,* in which he put forth the idea that the geological upheavals recorded in the Bible—such as the Great Flood, the Parting of the Red Sea, and Genesis itself—were caused by the earth's encounter with comets.

According to his theory, the earth was an ancient inert comet which, when struck by a second comet, began to spin in just the right way, so that "a paradise of life" soon developed on this newborn planet. But eventually, because of the grievous sins of humanity, God caused another comet to strike the earth and "inflict an awful punishment on man for his sins": The tail of a prodigious comet wrapped itself around the earth and caused the oceans to swell up and drown the planet of Noah's time "in a glorious religious purge." Whiston relates this Great Flood in detail:

On Friday, 28 November 2349 (BC) or on 2 December 2926 (BC), the comet situated at its node, and cutting the plane of the Earth's orbit at a point from which our globe was separated by a distance of only 3614 leagues of twenty-five to a degree. The conjunction took place at the hour of noon under the meridian of Pekin, where Noah, it appears, was dwelling before the flood.

Looking back, it's hard to believe such a theory could gain immediate and overwhelming acclaim among the scientific community. Yet it was regarded as "the noblest production of genius and science that had ever been given the world." In fact, for many years it was considered a more important document than Newton's *Principia,* which had been published a few years earlier. Voltaire later said that "Whiston was unreasonable enough to be astonished that some people laughed at him." But at the time not many were laughing. When the planet Mars made a close approach to Earth in 1719, comet fever again swept the world—many people believed it was Whiston's Great Comet on its way to destroy the world.

In our century, Whiston's ideas were restated with slight modifications by Immanuel Velikovsky, a Russian-born psychiatrist. He published *Worlds in Collision* and several other books that eventually caused a furor among scientists because so much attention was accorded Velikovsky's half-baked ideas.

To Velikovsky, the time scale of the universe was measured in thousands of years rather than in millions. He thus explained how within human history the giant planet Jupiter suffered a tremendous outburst and shot out a comet that later became the planet Venus. He wrote that "comet Venus" passed by the earth in 2500 B.C. at the time of the Israelites' exodus from Egypt and slowed the earth's rotation, drying up the Red Sea at a convenient moment for the Israelites to cross. Tremendous upheavals followed. Petrol rained down, so that our modern fuel is really "remnants of the intruding star which poured forth fire and sticky vapour." Two months later, after the earth had started spinning again, the "comet Venus" returned for a second visit, producing the thunder and lightning the Israelites saw when Moses received the Ten Commandments on Mount Sinai.

Other comet–Earth encounters followed, Velikovsky believed, causing, among other historical events, the tremors that shook down the walls of Jericho. The comet eventually collided with Mars, lost its tail, and subsequently became the planet Venus.

Of course, no one with even the most elementary knowledge of science could take Velikovsky seriously. In spite of this he developed a huge cult of followers, and when he died in 1979 his passing was mourned by a band of devotees numbering in the thousands.

CHAPTER 3

WHAT IF A COMET STRUCK THE EARTH? AND OTHER PARANOIAS

Every time a bright comet appears the fear that it will collide with Earth is renewed.

One of the most famous bouts of comet paranoia occurred in France in 1773, when Joseph Jérôme de Lalande, the mathematician, published a paper entitled *Reflections on the New Approach of a Comet to the Earth.* This certainly sounds innocuous enough. But suddenly word spread throughout the populace that a great fiery comet was about to strike the earth.

When Lalande got wind of the panic, he wrote a hasty disclaimer of any imminent comet–Earth collision. All, of course, to no avail. Human beings seem to need disaster to keep the blood flowing. The Archbishop of Rheims was called upon to provide a 40-hour prayer in hopes of somehow deflecting the onrushing calamity. And an enterprising clergyman sold "seats in paradise" to a number of wealthy fools. The comet missed the earth by well over 40-million miles.

A Dr. James Bedford warned the public of the dire consequences of a comet's headlong approach in his 1857 pamphlet, *Will the Great Comet Now Rapidly Approaching Strike the Earth?* This is written in a volatile, highly charged, pseudoscientific style that has a comical ring to 20th-century ears, but it was devoured by a very gullible public who bought up thousands of copies.

This all sounds like a lot of needless hysteria, yet a comet did collide with the earth in 1908 in a remote part of the Siberian plain.

28

Apparently what was described as a "fireball" appeared without warning and became "as bright as the sun." The engineer driving the Trans-Siberian Express more than 400 miles away from impact felt such a violent shock he thought his own train had exploded. When he stopped the train, his passengers told him they'd seen a bright blue ball of fire streaking across the sky trailing a tremendous tail of smoke.

At its epicenter, in the Podkamennaya-Tunguska River Valley, the blast flattened huge fir trees over 400 square miles of forest, wiped out most of the reindeer population of the area, and scattered the tents of nomads several hundred miles from the impact site.

Although this disaster has been blamed on everything from a flying saucer to a black hole, the evidence has eventually led scientists to agree that a very small comet did indeed strike the earth. In 1975 Ari Ben-Menahem, an Israeli scientist, reassessed all the known information and concluded that the main explosion occurred 8.5 kilometers above the ground and was equivalent to 12.5-million tons of high explosives or a moderately large hydrogen bomb. Another scientist, David Hughes, has calculated that the impact of a comet merely 40 meters in diameter and weighing about 50,000 tons would be enough to cause such a blast.

The Tunguska comet, in fact, coincided with a daylight meteor shower consisting of dust particles left in the orbit of the Comet Encke; so probably it was a small fragment of that comet. (The entire nucleus of Encke, itself only a medium-sized comet, was 100,000 times larger than the tiny comet that devastated a 400-square-mile area of Siberia.)

But what if a large comet struck a populated area of the earth?

In simple terms, if the comet were large enough and it landed, say, dead center in Manhattan, it would simply blow New York City off its islands and destroy large parts of New Jersey, Pennsylvania, upstate New York, and New England in the process. It's doubtful that it could jolt Earth out of its orbit, but it would be many times more devastating than the most powerful nuclear bomb yet developed and at least a million times more destructive than the bomb that killed 100,000 people and destroyed the city of Hiroshima in 1945.

The blast wave from such an impact would send a mushroom cloud of debris miles into the air, burst the lungs of anyone within 50 miles, and start a series of earthquakes throughout the world. If the comet landed in the ocean, enormous tidal waves would race across all the oceans of the world and destroy thousands of miles of coastal land and anyone unfortunate enough to be caught there.

The most devastating effect of a large comet's colliding with our

Earth, however, would be not the initial explosion but the calamities that would ensue for years afterward. The cloud of dust and debris from the impact would bring immediate darkness to the surrounding area. Farther away, grains spread on the stratospheric winds would at first produce murky sunsets and sunrises and turn the moon blue. A week or so later, the sun would disappear in a band around the world on approximately the same latitude as the explosion. Within a year, the entire Earth would suffer perpetual darkness.

In a phenomenon known as the Krakatoa Effect (after the volcanic island that erupted in 1883 and spread a thin layer of dust around the world for two years), a large comet, on striking the earth, would produce a mushroom-shaped cloud that would completely block all sunlight from our planet for at least four years. There is strong evidence to support the theory that a large comet did strike the earth 65-million years ago and that the ensuing death of almost all vegetation caused the extinction of the dinosaurs. Any who survived the blast skulked around in darkness for a while and slowly starved to death.

That's not all. There is evidence that a similar collision occurred 250-million years ago, driving the famous trilobites and most of the early reptile species into extinction. Also, a less severe collision 365-million years ago doubtless caused the disappearance of many marine invertebrates. So the possibility of a large comet's striking the earth is very real indeed.

Comet–Earth collisions, in fact, have always occurred. Unlike our pockmarked moon, which shows the effects of its millions of encounters with comets, apollos and meteors, though, the earth does a pretty good job of hiding its scars with vegetation, snow, and oceans of water.

Before you rush out to buy a ticket on the space shuttle, you should know that the odds *against* a comet's striking the earth are more than a billion to one—which works out to about once every 80-million years. You should worry more about crossing the street on your way to pick up your ticket.

If you're smart, however, and human behavior sticks to form, it might prove wise to *invest* in the space shuttle. Then, next time a large comet threatens a close pass to the earth, the ensuing panic could set you up for life financially.

PART TWO

THE STORY OF EDMUND HALLEY AND HIS COMET

Halley and his comet.

CHAPTER 1

THE UNIVERSE BEFORE NEWTON

It was Sir Edmund Halley, in collaboration with no less a figure than Sir Isaac Newton, who put comets on the map. As far back as 1682, when he first saw the Great Comet of that year, Halley had the idea that would make him famous forever: That was when he first began to believe that comets, like the planets, return in regular periods to the sun. That was a revolutionary idea. Before Halley, comets were thought to travel in straight lines—never to be seen again once they passed us by.

To prove his assertion, however, he had to answer the fundamental question many scientists were asking: What force held the universe together?

It was now almost universally accepted that the planets revolved around the sun. But what was the cause of it all? A lot of educated guesses were flying about, but no one had the power or breadth of mind to bring them all together.

Fortunately, Isaac Newton was on the verge of making those discoveries that would bridge the knowledge gaps. But he needed help. And Halley, realizing Newton could open the way to the greatest scientific revolution the world had yet seen, spared no effort and no expense to see that Newton's theories came to light.

By the time Isaac Newton was born, it had been established that the sun rather than the earth is at the center of our Solar System. The men chiefly responsible for this change of view were Copernicus, Kepler, and Galileo.

Aristotelian physics ruled the world for nearly two thousand years. It was Ptolemy who had synthesized Aristotle's much-borrowed thoughts into a cogent mathematical treatise in the second century A.D., in his great treatise, *Almagest.*

Ptolemy constructed a theory and a mathematics for what anyone could plainly see—the sun and the moon and the planets moved around the earth. The sun rose in the east and went down in the west. And to agree with the classical idea of the perfection of the universe, held since Pythagoras in the 6th century B.C., Ptolemy stated that these heavenly bodies all traveled in great circular paths around our Earth with a steady, regular motion.

But there was at least one major flaw in this concept of the perfect universe moving in perfect circles. No one could convince the planets to cooperate. They wandered irrationally all over the sky, weaving in and out among the fixed positions of the stars. Sometimes they would even stop and "reverse" direction! Quite a few geniuses were stumped for quite a few centuries over that one. Finally, a theory known as "the deferent and epicycle" emerged, and that's the way Ptolemy explained the disorder within the heavenly order. Simply stated, a planet was thought to move in a small circle (the epicycle), whose center moves at the same time around the circumference of a larger circle (the deferent). This theory led over the centuries to the need for a whole truckload of epicycles to be added and subtracted from the basic picture in an attempt to diagram all the strange behavior in the sky. It was all a beautifully elaborate clockwork of different-sized circles, and circles within circles, and circles moving in syncopation with other circles. And, of course, it was all dead wrong.

This perfect world began to falter sometime around the 13th century. But it was not until 1543, when Nicolas Copernicus's monumental work, *De Revolutionibus Orbium Cœlestium (On the Revolution of the Celestial Spheres),* was published, that the Ptolemaic concept collapsed completely.

Copernicus, of course, put the sun at the center of our Solar System, with Earth and the other planets revolving around it. He also explained that Earth rotated once a day on its axis and made a complete revolution around the sun every year. The apparently erratic nature of the motions of the other planets could then be explained as merely illusion created by the real movements of Earth.

Copernicus did stick to one Aristotelian theory, however: that the immutable universe is composed of perfect spheres moving in perfect circles. This little bit of perfection messed up the whole thing. The planets still deviated from the positions where circular paths *should* have taken them. So Copernicus, too, had to use epicycles to account for these deviations.

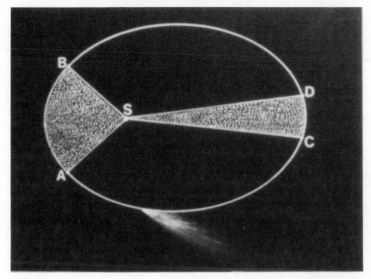

Kepler's Second Law: The radius vector of an ellipse will sweep out equal areas in equal time. Area ABS is equal to area CDS. So the comet or planet will travel from A to B in the same length of time it will take to travel from C to D. In simple terms, this means that a body in orbit will travel faster when it is closer to the sun.

It took Johannes Kepler, with the help of the great observational astronomer Tycho Brahe, to set things straight. Kepler, analyzing Tycho's detailed observations, concluded that Mars traveled around the sun not in a circle but in an ellipse. That is, its orbit was shaped more like a football than a basketball. Kepler quickly generalized this theory to fit all the planets and, in 1609, published his *Astronomia Nova (New Astronomy),* which describes his new "laws" of planetary motion.

Kepler's explanation that planets indeed have elliptical orbits is known as Kepler's First Law of Planetary Motion. With it Kepler, in one clean sweep, eliminated all need for epicycles, eccentrics, and all the incredible geared machinery that was thought for centuries to control the movement of the planets. All the deviations and questions raised by the "wandering" paths of the planets were answered by the elongated, curved line of the ellipse.

Kepler's Second Law of Planetary Motion describes how a body in an elliptical orbit travels faster as it gets closer to the sun: A line drawn from the sun's focus to the planet moves at a rate that causes it

35

to "sweep out" equal areas inside the ellipse in equal times (see diagram). This meant that the speed and, therefore, the position of the planet at any moment could now be computed.

In his *Harmonices Mundi (Harmonies of the World)*, published ten years later, Kepler added a third law, which tied the planets together into an organized system by showing that the distance each planet is from the sun is related to the length of time each takes to complete its orbit.

Kepler's laws fit with the physical evidence known at that time, as they were derived mostly from Brahe's remarkably accurate observations. But Kepler could not explain *why* the planets moved in these elliptical paths.

And that's where Newton comes in.

CHAPTER 2

THE WORLD ACCORDING TO NEWTON

When Newton was born, the heliocentric universe was an accepted fact. The ideas of Copernicus and Kepler had had nearly a century to be thought and rethought, hashed and rehashed, by those minds that counted and had finally percolated down to the common man. But the matter was still far from settled. Kepler's theories left as many questions as answers. Aristotle and Ptolemy's homocentric spheres gave man an understandable physical picture of the heavens—something he could grasp, something he could visualize and understand.

Tycho Brahe's laborious observations of the sky had been accurate enough to explode the "perfect spheres" idea of Aristotle. But with these spheres that supposedly moved about on tracks of perfect circles in the majestic heavens went man's understanding of the workings of the universe. Kepler's so-called New Astronomy suggested that the planets traced and retraced well-defined ellipses through a trackless space without any guidance or structural support. So scientists of the 16th and 17th centuries were led to the inevitable question: What force holds the Solar System—and, in fact, the entire universe—together?

In 1684, a young scientist named Edmund Halley was pondering this same question. At a meeting of Britain's Royal Society he discussed the problem with two of the most prominent scientists of the day, Sir Christopher Wren and Sir Robert Hooke. It was a feisty, good-natured discussion, attended by a little alcohol and a lot of scholarly braggadocio.

Sir Isaac Newton

Hooke, who was president of the Royal Society and would later take center stage in a drama concerning the originality of Isaac Newton's theories, was a man of immense self-importance. He claimed without hesitation that he had hit upon the answer to the problems of celestial motion. Now Halley, Wren, Hooke, and a few others already generally agreed that, by extrapolation of Kepler's

Laws, the laws of heavenly motion must follow the inverse-square relation. That is, the farther from the sun, the less the pull of gravity on the planets. But proving the theory involved a considerable mathematical challenge. That didn't daunt Hooke, however. He told Halley and Wren that he, indeed, had the answer but was only keeping it from the world in order that other men, trying and failing, could see just how difficult and valuable his calculations were.

Wren, in a friendly mood, offered a prize to whoever brought him a demonstration of the inverse-square principle within two months.

Hooke never delivered.

Halley was not hampered by inflated pride and had no delusions that he could crack the problem—but he did have a pretty good idea who could. In August he went to visit Newton at home in Cambridge. That meeting would forever change the way Mankind viewed the world.

Halley was a good-natured, well-liked, gregarious man of 28 when he went to visit Newton for the first time. Newton, in contrast, was a moody, sometimes arrogant, sometimes paranoid, cranky, reclusive man in his 40s. He had by then a formidable reputation as a man of science and a man of dark moods and he could be difficult indeed.

Newton had had an awkward childhood. His father had abandoned him and his mother. This had left him confused and violently bitter and overly dependent on his mother. He never married and he had few close friends. He viewed other scientists not as colleagues but as rivals, and he was reluctant to publish his mind's overflowing ideas.

Halley, however, had immense charm and he greatly admired Newton. Also, he could spot a genius when he saw one, and he never let Newton's great accomplishments hurt his own pride: He only cared that truth and the secrets of the universe were being revealed. And another thing: He genuinely liked Newton—which in itself was a tall order.

It speaks volumes for Halley's character that although just about every other great scientist of the time fell victim to raging fits of jealousy over Newton, Halley did not. In fact, the moment he realized just how monumental Newton's genius was, he embraced Newton and did all he could to assist in the birth of his ideas. In doing so, Halley played midwife to some of the greatest scientific discoveries of any age.

But Halley, even at this early age, was no slouch himself. While only 18 years old and a student at Oxford, he had attempted to improve the astronomical tables for Jupiter and Saturn, but at once he saw that, if he were to accomplish this task, he would need a more accurate catalog of the fixed stars than was currently available. So he had packed his telescopes and clothes and set sail for the island of St. Helena, in the South Atlantic. Two of the world's leading astronomers of the time, J. Hevelius and John Flamsteed, were at work on star catalogs of the northern hemisphere, and Halley meant to put the southern stars on the map and thus add to the great work. He stayed on St. Helena nearly two years, and when upon his return he published the results, his *Catalogue of the Southern Stars* was acclaimed a great contribution to astronomical science. Halley had been named a Master of Arts by Oxford for this work. Then, in November of 1678, he had been elected a Fellow of the Royal Society in recognition of his work. He was then 22.

When Halley went to visit Newton in 1684, the two men hit it off immediately. Newton trusted Halley from the start, and when the younger man asked Newton if he could prove the law of inverse-square, he revealed without hesitation that he could.

Halley, of course, was overjoyed. But there was a problem. Newton had mislaid his calculations somewhere in the voluminous mess of his office and couldn't find them. But he promised to work out all the calculations again and send them to Halley as soon as possible.

Unlike Hooke, Newton was good to his word. Newton's mathematical proofs of Kepler's concepts came by messenger to Halley in London soon after their meeting, and at once Halley knew he had something special in his hands.

What Halley received was a nine-page treatise titled *De Motu Corporum in Gyrum (On the Motion of Bodies in an Orbit)*. In it, Newton not only demonstrated mathematically Kepler's First Law of Motion—the inverse-square principle—but proved Kepler's Second and Third Laws, as well.

Halley knew at his first inspection of *De Motu* that Newton had taken such a leap forward in understanding celestial motion that a revolution was at hand: Newton was about to explain what made the universe work.

Halley dropped everything he was working on and went immediately to Cambridge to see Newton again. He wanted to persuade Newton to begin at once to set down in a full volume this work of such immense importance. He knew that Newton had let these ideas moulder among the stacks of notes in his office, and he wanted to see that the different pieces of this tapestry were pieced together into one great picture of the universe.

Halley succeeded in this ambition. *Philosophiae Naturalis Principia Mathematica (Mathematical Principles of Natural Philosophy)*, or the *Principia,* would finally be published in 1687, but first it was introduced to the Royal Society. In it, Newton proved that the elliptical paths of the planets orbiting the sun, first determined by Kepler, were the consequence of the law of gravity.

As the Royal Society recorded on April 28, 1686:

> Dr. Vincent presented to the Society a manuscript treatise, entitled *Philosophiae Naturalis Principia Mathematica,* and dedicated to the Society by Mr. Isaac Newton, wherein he gives a mathematical demonstration of the Copernican hypothesis as proposed by Kepler, and worked out all the phænomena of the celestial motions by the only supposition of a gravitation towards the center of the sun increasing as the squares of the distances therefrom reciprocally.

All well and good. But how was it that Newton had decided that gravity was the force that bound it all together?

As it turns out, the story of the apple that fell and hit Newton on the noggin and sparked this genius into thought was not all fable. Although the account of the incident by his assistant, Henry Pemberton, written when Newton was an old man, leaves out the apple, many others have stated that Newton indeed confirmed that it was an apple that set off his ruminations.

> As he sat alone in a garden, he fell into a speculation on the power of gravity: that as this power is not found sensibly diminished at the remotest distance from the center of the earth, to which we can rise, neither at the tops of the loftiest buildings, nor even on the summits of the highest mountains; it appeared to him reasonable to conclude, that this power must extend much farther than was usually thought; why not as high as the moon, said he to himself? and if so, her motion must be influenced by it; perhaps she is retained in her orbit thereby. However, though the power of gravity is not sensibly weakened in the little change of distance, at which we can place our selves from the center of the earth; yet it is very possible, that so high as the moon this power may differ much in strength from what it is here. To make an estimate, what might be the degree of this diminution, he considered with himself, that if the moon be retained in her orbit by the force of gravity, so doubly the primary planets are carried round the sun by the like power. And by comparing the periods

of the several planets with their distances from the sun, he found, that if any power like gravity held them in their courses, its strength must decrease in the duplicate proportion of the increase of distance.[1]

If gravity could make that apple drop to the ground and, also, extend its influence to the highest mountain, then, why couldn't it reach to the moon and the planets and, thus, keep them in their orbits? Why not, indeed?

Newton concluded—rightly, as we now know—that this force that keeps the planets in their orbits varies inversely as the square of the distance from the sun, thus agreeing with Kepler's Third Law.

Newton had figured out the crucial point that what was needed in order to keep the planets in their orbits was not a force pushing them along from behind (as in the French philosopher René Descartes's idea of Vortices) but a force pulling them *in toward* the sun—so that, with the speed of their motion, they did not simply shoot in straight lines off into space but instead were swung around the sun, thereby creating an orbit.

But Newton took the laws of gravity even further. He was convinced that Kepler's tables concerning Saturn's motion had been wrong. He reasoned that, if gravity was a force that applied to all bodies, not just the sun pulling upon the planets, then when Saturn and Jupiter passed close to one another their paths would be altered.

Therefore he asked John Flamsteed, who was the Royal Astronomer at the time, for his observations of the velocities of Saturn and Jupiter as they neared each other in their respective orbits. If Newton was right, Saturn should slow down in its orbit as it approached Jupiter and speed up as it passed. And those were exactly the results Flamsteed reported to Newton. Newton was ecstatic. He now knew he had discovered not just the force that held the Solar System together but the single universal attraction that made the entire world go around. And that of course included comets.

Newton wrote to Flamsteed again to obtain information about tidal movements in the estuary of the river Thames, and again he was confirmed in his suspicion that gravity influenced every body, no matter how small—even a body of water.

In the *Principia* there appeared for the first time a system of laws that tied together the movements of celestial bodies with the way bodies moved on Earth: With, for example, an apple falling from a tree. This was an idea of immense breadth that explained all the physical universe. It was a concept that the world had long needed.

Without Edmund Halley, the *Principia* would almost certainly not have been published. Augustus De Morgan said, "But for him, in all human probability, the work would not have been thought of, nor when thought of written, nor when written printed."

In fact, it was Halley's idea to extend the theories Newton expressed in the *De Motu* into a vast treatise enveloping all the physical phenomena of our Solar System—the motions of the planets, the moon, and even comets. But getting that work published turned out to be a long, arduous, frustrating battle. One problem after another, one person after another tried, it seemed, to stop the *Principia* from being published. But Halley let nothing stop him—not even Newton.

In the first place, Newton himself did his best to suppress his work. He was reluctant to publish his ideas because he felt others would try to steal them. As it turned out, he was justified in his fears.

Secondly, celestial mechanics was not Newton's real passion. He was enthralled by alchemy. That was the mysterious frontier of the intellect and spirit that dominated his hungry mind for years until his other, greater scientific ideas were in danger of becoming lost among a shuffle of papers.

From the time Halley persuaded Newton how important his ideas were and that he should undertake the laborious task of publication, it took Newton nearly two years of constant work to complete the manuscript. Newton had revealed a great truth of nature, and that truth sent out a beam of light so strong it illuminated question after question that had puzzled man until that moment. Newton had in his possession a great lantern, and the farther he followed its light, the more it revealed.

The task was monumental. But once Newton got a problem in his jaws, he was relentless. He spent almost his entire waking and sleeping hours in his laboratory at Cambridge, accomplishing very little else and having almost no social contact while he worked on the manuscript.

There was always something or someone, it seemed—some misunderstanding, some misfortune, or some missing ingredient—that slowed Newton and Halley's progress, however.

In 1684 tragedy struck Halley, and the consequences of the incident would upset his life and his work for years to come. In March his father's bludgeoned body was found on the bank of the Medway River near Strood. The details surrounding his father's death were mysterious. The whole thing was an odd, inexplicable puzzle.

Edmund Halley senior was a rich and prosperous man, and that surely was his undoing. He made his money mostly from a soap-boiling and salting business, and these endeavors had paid off big when bubonic plague ravaged London and all of Europe between 1664 and 1666. People were understandably fanatic about cleanliness at this time. For example, merchants refused to touch money from customers but, instead, had them drop it into a pot of vinegar. So, while many businesses were failing, Halley's soap business boomed. He owned a townhouse and business in Winchester Street in the City of London, as well as several properties that netted him more than a thousand pounds a year in rents. He also owned a country home in Hackney, three miles northeast of St. Paul's Cathedral a good distance from the noise and filth and soot of downtown London.

Edmund Halley's father, then, had been a happy and successful man. But things suddenly seemed to go all wrong when, in his later years, he married for the second time. His new wife was many years younger, a pretty, clever woman who knew how to spend her husband's money and who, once installed in his home, showed little affection to either the elder Halley or his son.

The circumstances of his death were baffling. He went out one night, telling his wife he would be back in a few hours. When he did not return, she went the next day to the police to report him missing, and four days later she offered a reward of £100 to anyone who found him—dead or alive.

Five days after Halley left his home, his body was discovered by the side of the river Thames at a place called Temple Farm. *The London Gazet* reported the discovery:

A poor Boy walking by the Water-side upon some Occasion spied the Body of a Man dead and Stript, with only his Shoes and Stockings on, upon which he presently made a discovery of it to some others, which coming to the knowledge of a Gentleman, who had read the Advertisement in the Gazet, he immediately came up to London, and acquainted Mrs Halley with it, withal telling her, that what he had done, was not for the sake of the Reward, but upon principles more Honourable and Christian, for as to the mony he desired to make no advantage of it, but that it might be given entirely to the poor Boy; who found him and justly deserved it.

The body was badly mutilated, the face bashed in, and, since it had been in the river at least four days, horribly bloated. It was only because the soles and linings of the elder Halley's shoes had been distinctly altered that the family were able to identify his remains.

Halley at about ages 25, 45 and 65.

Murder, of course, was immediately suspected. But there were those who, because of some odd things about his death, thought he had committed suicide. For example, Halley was found stripped of all clothes except *four* pairs of stockings that he was wearing. He had also been thought to be suffering from "private discontent" over his marriage. But, in the end, the jury's verdict was murder: It was concluded that the murder was committed elsewhere and the body dumped in the river later. His murderer was never found.

The grief over his father's death not only slowed down the younger Edmund's collaborative efforts on the *Principia,* but it was a serious financial burden, as well. Halley's father had always been most generous with his son and supported him in his efforts; so Halley had always been able to pursue his studies and research without a great deal of worry over where his next meal would appear from. Upon his father's death, however, not only was his income cut short but he also had to battle his stepmother for his inheritance. She married again soon after, and Halley and she were still embroiled in court actions ten years after his father's death.

The problems with the *Principia* continued. In May 1686 the Royal Society agreed to publish Newton's book and appointed Halley to oversee the editing and printing. But a month later their decision was reversed, and thus the Royal Society declined to print one of the greatest books of science ever written.

Halley, whose monetary fortunes were at a low ebb because of his father's death, dug into his own pocket to have the book published. He realized the enormous range of Newton's theories and their magnificent contribution to the understanding of our world. And he put his money where his beliefs were.

CHAPTER 3

A STORM OF JEALOUSY

The publication of the *Principia* took by storm those in the learned world who were able to understand it. The world fell at Newton's feet. He was recognized almost universally and immediately as a man of immense genius, and his work as perhaps the single greatest scientific achievement yet seen by Mankind. But there was a problem created by the collection of powerful personalities that made up Britain's Royal Society in the 17th century. When Flamsteed, Wren, Hooke, Whiston, Newton, and others met together, there was an atmosphere like a roomful of divas. Halley was an easygoing, affable man who had a warm affection for nearly everyone he knew. But the temperamental jealousies and petty bickering that ran through the Royal Society embroiled even him in its tempestuous atmosphere.

Keep in mind that these were all men of extraordinary talent and achievements who, each in his own right, had blazed new trails of thought and scientific discovery. Then along came a man who at once outshone even the brightest among them and made their greatest advancements seem like infants' steps.

The man most torn by jealousy was Robert Hooke. Immediately after the final presentation of Newton's *Principia* to the Royal Society on April 28, 1686, Hooke publicly accused Newton of plagiarism. Newton, Hooke insisted, had first learned of the inverse-square law of gravitation from him, and thus had stolen the idea! All Newton did was work up the mathematical formula for his discovery, Hooke said; and Hooke wanted Newton to acknowledge his contribution.

Hooke and Newton had tangled before (in 1672, over Newton's early theories about light) and the bitter rivalry had never ceased.

They were the perfect rivals—a matched set of opposites. No two men were more different and, because both were stubborn and narrow-minded in their own ways, they clashed from almost the moment they met.

Hooke was a genius—the leading scientist of his day before Newton and his prodigious talent encompassed all fields of knowledge, including physics, chemistry, physiology, and biology. He made great strides toward explaining the nature of light, combustion, animal respiration, and the structure of cells. He also invented a diving bell and a method of telegraphy, and refined watch gears.

Hooke had ferocious energy, and he skipped from project to project, dabbling in whatever interested him at the time. To the somber Newton, Hooke's sort of scientific theorizing was bombastic—and it reflected his life.

Hooke extolled creature comforts and the sensual life, and he spent a lot of time socializing and lionizing in taverns, half-drunk and well on his way to dead drunk. He was witty, showy, powerful—and he knew it. He had affairs with a string of his housekeepers at Gresham College in London, where he was Professor of Geometry, and he didn't care who found out about it. When his niece, the daughter of his deceased brother, came to live with him, he promptly fell in love with her. Hooke, the middle-aged romantic, blew this into his "grand passion," and he agonized over and sentimentalized her unfaithfulness to him.

Hooke was all excess in everything he did. This irritated Newton, the staunch Puritan who rarely smiled and never loved any woman but his mother. Newton was grouchy, revengeful, sardonic, and almost joyless. In short, Newton could be a monumental pill. But he had one of the greatest minds ever seen on this planet and he loved his work as only a fanatic can love.

Halley was one of the first to hear of Hooke's claim, and he feared Newton's reaction. All Halley wanted was to shepherd the great truths that Newton held in his grasp into publication. He knew of Newton's paranoia, though, and feared his reaction would be to keep the world from sharing his discoveries. His letter to Newton upon hearing of Hooke's accusations reflects his fears:

> There is one thing I ought to inform you of, viz, that Mr Hook has some pretensions upon the invention of the rule of the decrease of Gravity. He sais you had the notion from him, though he Owns the Demonstration of the Curves generated thereby to be wholly your own; how much of this is so, you know best, as

likewise what you have to do in this matter, only Mr Hook seems to expect you should make some mention of him, in the preface, which, it is possible, you may see reason to praefix. I must beg your pardon that it is I, that send you this account, but I thought it my duty to let you know it, that so you may act accordingly; being in myself fully satisfied, that nothing but the greatest Candour imaginable, is to be expected from a person, who of all men has the least need to borrow reputation.[2]

Newton replied calmly to Halley, thanking him for informing him of Hooke's claims. Then, he lost his temper. It seemed to him Hooke was trying to reduce his work to simply a mathematician's finishing touches on Hooke's great idea.

Newton's denial was swift and brutal. He was prepared to go to any lengths, however childish, to defy Hooke's claims. He threatened to suppress the publication of the entire third book of the *Principia*. He insisted Hooke could lay claim to none of the propositions laid out in the *Principia*.

In all fairness, the ideas of gravity as they applied to Kepler's inverse-square law had been discussed seriously for more than a decade. But neither Hooke (who, according to a recent Newton biographer, Richard S. Westfall, never had a theory of universal gravity, but merely asserted the existence of particular gravities and that his inverse-square relation was a "medley of confusion") nor Newton had really discovered gravity. The idea had been kicked around by a lot of minds for a lot of years.

But neither Newton nor Hooke would compromise.

Halley was caught in the middle. He had to keep these two great intellectual Titans, these two emotional infants with minds like gods and all the social and political power of petty potentates, from crushing this great truth before it was fully born.

It was not as if Halley were suited for the job, because Halley was not by nature a great diplomat. In fact, he was sort of, well, rowdy. He also didn't mind breaking tradition and propriety and, whenever he got the chance, thoroughly disrupting the stuffy Puritan strain of the Royal Society and its gatherings. (There is a famous story that all Halley biographers tell concerning the time when Czar Peter of Russia visited London in 1698. Peter was 26 at the time and had a bold, bumptious streak equal to Halley's. He lost no time seeking Halley's company. They became immediate friends and spent many nights in a tavern on Great Tower street drinking ale and smoking until the wee hours. Legend has it that late one evening the two inflicted great damage to the gardens at Sayes Court by wheeling each other in a wheelbarrow *through* the hedges.)

But even though Halley tweaked a few noses, he was still well liked and got away with his outspoken behavior because he was so popular. It also helped, of course, when he became the greatest astronomer of his age. Yet later, when he got too candid about his religious beliefs, which were no beliefs at all—particularly that Puritanism was a big bore—he got himself in trouble: He asserted that perhaps not all of scripture was strictly true, and that was enough to set off a furor in the Royal Society and to get him charged with heresy under the Act of Uniformity. Although he was never actually punished as a criminal, he was denied the Savilian Chair of Astronomy, which he was in line for.

But there also may have been a little politics involved in that. There is a story that when it got out that a Scotsman named David Gregory had been given the position instead of Halley, another Scotsman, upon hearing of this, walked into the coffeehouse Halley was known to frequent, and demanded to see Halley because, he said, ''I would fain see a man that has less religion than Dr. Gregory.''

Halley never really cared about his own personal reputation, and he dismissed the small minds of those prigs who objected to his carefree lifestyle. But when in 1686 those small minds stood in the way of the greatest scientific mind the world had seen to date, he fought and compromised and did all he could to soothe the bruised feelings of those who opposed Newton.

Upon hearing of Newton's plans to suppress the publication of the third book of the *Principia,* Halley went immediately to Christopher Wren to sort out the mess. From Wren, Halley learned that Hooke had tried many times within the previous few years to solve the inverse-square problem mathematically, but his computations all eventually led to blind alleys. Wren admitted that he himself had also tried—and failed—to solve the problem.

It was obvious to Halley, then, that Hooke had only *raised the question* of gravity, just as many others had. But Newton had *answered* all these questions and a lot more.

Halley then approached Hooke and ''plainly told him, that unless he produce another differing demonstration, and let the world judge it, neither I nor anyone else can believe it.''

In this matter, the Royal Society eventually backed Halley. Hooke formally raised his objections to Newton's plagiarism, but his claims were soundly dismissed. This final rebuke from his peers at the Society crushed Hooke. It was a devastating calamity, from which he never really recovered either in reputation or in spirit. Another recent biographer explained Hooke's demise:

When two powerful figures challenged each other in the small, cramped, scientific societies of the seventeenth century, the clash resounded. There was not room at the head of the Royal Society for both Hooke and Newton. In an almost animal sense one and one only could be the champion of knowledge. Newton was an intruder into a kingdom where Hooke had once reigned supreme, and as the new star rose, Hooke lost power and status. He fought for his position, belittling the achievements of the newcomer, repeatedly charging him with plagiarism, clinging tenaciously to his post until his death in 1703—toward the end mere skin and bones. It was only then that Newton assumed the presidency of the Society and became uncontested leader of the herd.[3]

It has been said of Newton that 20 years after Hooke's death he could still not speak his name without losing his composure. It was a bitter hatred that brewed in these two men over the discovery of gravity.

Curiously, even though it was Halley who had called Hooke's bluff on the question of Newton's theories, both Hooke *and* Newton remained loyal and close to Halley until their deaths.

Without Halley, Newton would never have written the *Principia,* and Newton, in a rare moment of humility, acknowledged Halley's part. He wrote in the Preface (dated May 8, 1686):

> In the publication of this work the most acute and universally learned Mr. Edmund Halley not only assisted me in correcting the errors of the press and preparing the geometrical figures, but it was through his solicitations that it came to be published; for when he had obtained of me my demonstrations of the figure of the celestial orbits, he continually pressed me to communicate the same to the Royal Society.

Now that Halley had seen the *Principia* to completion, he held the key to answering a question he had been struggling with for years.

Halley's Comet was about to become famous.

CHAPTER 4

COMETS BEFORE HALLEY

The often spectacular nature of comets, their mysterious origins and their unpredictable appearances, have made them fertile objects of scientific speculation and argument since the ancient Babylonians. At first the main debate centered on whether or not comets really were celestial objects at all, or merely disturbances of our own atmosphere.

It was the Babylonians who first proposed the theory that comets, like planets, orbited the sun but that comets appeared only infrequently because their orbits were very much larger than those of the planets. That is substantially what we know to be true today. The Greek philosophers of the Pythagorean school of the sixth century B.C. and Hippocrates a century later concurred with the Babylonians, adding that the tail was an insubstantial "illusion" created by reflection. Unfortunately, almost no one else for several thousand years agreed with them.

For example, Democritus, around 420 B.C., thought that when comets dissolved they left stars behind. The Greek historian Ephorus had the opposite theory: He thought comets were made when two stars converged. He based his hypothesis on the observation of the comet of 371 B.C., which he said "split" into two separate stars.

Several hundred years later Seneca, in the first century A.D., reviewed comet history and was outraged at Ephorus's observations. He thought Ephorus had been grandstanding in order to gain public attention. Ironically, we know today that comets often *do* split into two or more separate comets, although they don't have a thing to do with stars.

It was Aristotle's ideas, as presented in his fourth century B.C. treatise on meteorology, that dominated the science and astronomy of western civilization for nearly two thousand years. Aristotle disputed the periodic nature of comets because they had been observed outside the zodiac, the domain in which the planets were known to wander. He felt sure they did not act as the planets did.

Aristotle's most potent argument, and one that held the most force for thousands of years, was his concept of the immutable, perfect universe—a heavenly perfection that moved with godlike precision. Comets were anything but precise. They were irregular and unpredictable and often dissolved without a trace. Aristotle therefore concluded that comets were not heavenly bodies but rather disturbances of our upper atmosphere occurring well below the moon. He figured that they were "hot, dry exhalations" rising from the ground and carried along by the motion of the sky. As they were pushed along they were heated by friction and eventually burst into flame. If the movement were slow, a comet was produced; if fast, a meteor. (He believed the Milky Way was created by the same material.) Also, following his own logic and the belief generally held at that time, Aristotle accepted the idea that comets contributed to droughts and high winds.

Seneca's was one of the few dissenting voices for nearly two thousand years. In his *Questiones Naturales,* Seneca picks apart Aristotle's opinions about comets one by one:

If it [a comet] were a wandering star (*i.e.*, a planet), says someone, it would be in the zodiac. Who, say I, ever thinks of placing a single bound to the stars? Or of cooping up the divine into narrow space? These very stars, which you suppose to be the only ones that move, have, as every one knows, orbits different one from another. Why, then, should there not be some stars that have a separate distinctive orbit far removed from them? There are many things whose existence we allow, but whose character we are still in ignorance of . . . why should we be surprised, then, that comets, so rare a thing in the universe, are not embraced under definite laws, or that their return is at long intervals? . . . The day will yet come when the progress of research through long ages will reveal to sight the mysteries of nature that are now concealed. The day will yet come when posterity will be amazed that we remained ignorant of things that will to them seem so plain.

Aside from Seneca's objections, however, the study of comets stood still for many centuries, and Aristotle's beliefs about comets were strengthened, particularly his belief that comets were omens of evil, death, and destruction.

The question of whether comets were or were not celestial objects was finally settled by the Danish astronomer Tycho Brahe. Tycho was a bombastic oddball who wore pretentious robes while observing and had his tenants tossed into jail when their rents went delinquent. Eventually he was thrown out of Denmark by the same royalty that had built him a great observatory on the island of Hven; but during the 20 years he worked on his island he made phenomenal leaps in observational astronomy.

He questioned Aristotle's theory that comets originate in our atmosphere because, if that were true, there seemed to be no reason why the tails always pointed away from the sun.

Tycho figured that the only conclusive proof would be to measure a comet's diurnal parallax. This is a change in the *apparent* position of an object in the sky, which is actually caused by the changing position of the observer as the earth rotates. Luck was with him, because in 1577 a comet bright enough to be seen in the daytime appeared over Europe. Tycho measured the position of the comet at its high and low points in the sky and found only an insignificant movement in its relative position. An atmospheric object would have an easily measurable parallax: If it were closer than the moon, its position would have shifted against the position of the stars as the earth rotated. But this had not happened.

Tycho then compared his results with those of Hagecius in Prague, 600 miles away, and again found a difference of only one to two minutes of arc. During the same time, the moon had moved significantly—which implied that the moon was far closer to Earth than the comet was. He therefore calculated that the comet was more than a million kilometers away, at least six times the distance to the moon.

Of course, there was the usual noisy dissent that breaks out when any discovery goes against prevailing belief. Even Galileo Galilei, who would play a major role in establishing the Copernican theory and was to be the first great practitioner of telescopic observation, disputed Tycho Brahe's conclusions. He stated that Brahe's observations could easily have been an optical illusion, and that comets were nothing more than the reflected sunlight of vapors that had risen beyond the arc of Earth's shadows.

For years after Tycho had finished his work on comets in 1596, many European universities made their professors sign documents renouncing the "false doctrine" that comets were heavenly bodies. But eventually, with the weight of the truth on his side, Tycho Brahe's conclusions were generally accepted.

Tycho Brahe's proof that comets were indeed celestial objects led to the second great question concerning comets: How do they move in the sky? through space in straight lines or in orbits around the sun or the earth?

To answer this question, a comet's path had to be charted. With the crude telescopes then in use, it was extremely difficult to map a comet's flight with any great accuracy. Even the largest comets were visible for only a few weeks to the naked eye and for only a few months to telescopes as they traveled through several constellations before fading away into the background of stars.

Brahe himself attempted to fix an orbit for the comet of 1597 that was nearly a circle. But his conclusion clashed with what he saw, and he had to admit that the comet's motion was actually ovoid. As J.L.E. Dreyer, the British astronomer, points out, "This is certainly the first time that an astronomer suggested that a celestial body might move in an orbit differing from a circle, without distinctly saying that a curve was the resultant of several circular motions."[4]

Just a short time later, in 1609, Kepler was to show that the planets' paths were, indeed, ellipses; but, curiously, he never believed comets described a similar orbit. Instead, he thought they followed straight paths at irregular speeds.

During the next 50 years, several scientists made attempts to dispute Kepler's conclusion, but without much success.

Hevelius of Danzig, for instance, believed Kepler to be right about the straight-line travel of comets until meticulous observation of four different comets contradicted that. Hevelius then concocted his own subtle and complicated theory, which stated that comets were disk-shaped objects expelled from the planets Jupiter and Saturn. As they approached the sun, these disks flipped over to one edge. This caused their path to curve slightly, thus creating a very wide arc.

Hevelius himself never suspected that the comets were actually *orbiting* the sun—only that they were influenced by it. It was a student of his, however, George Dorfel, who was later to suggest, while studying the comet of 1680, that comets moved in parabolas and that the focus of their path was, indeed, the sun.

They were getting closer now to discovering the way the world worked.

CHAPTER 5

HALLEY'S WORK ON COMETS

Once Halley finished the herculean task of getting the *Principia* out of the mind of Newton and down on paper and published, he wasted no time in using what he had learned.

Kepler's laws had really been meant to describe planetary motion. Now Newton had not only explained why the planets move as they do but also extended the theory to describe and explain all celestial movement—and that, of course, included comets.

If Newton were right about all this, then comets, following the same physical laws that govern the planets, might also revolve in elliptical fashion about the sun. From observation, all that could really be established was that comets approached the sun from somewhere in space and disappeared again back into space. According to Newton's physics, however, all celestial objects that passed through our Solar System were influenced by the sun's gravitational power. As a result, they could take only one of five paths: a straight death drop directly into the sun, a perfect circle, an ellipse, a parabola, or a hyperbola.

No one had ever seen a comet fall into the sun, and Kepler had proved that perfect circles were not the shape of celestial orbits. So that left ellipses, parabolas, and hyperbolas.

Ellipses are closed planes. That is, at some point (depending on the eccentricity or elongation of their curve) they come back full circle upon their original path, back to their original starting place, and begin all over again. *Parabolas* are, theoretically, elongated ellipses stretched out to infinity—something that rarely occurs in nature. A *hyperbola,* on the other hand, is an open plane. It never comes back full circle. If a comet moved in a hyperbolic path, it would make one pass of the sun and never return.

The law of gravity as it applies to orbiting bodies is the Catch-22 of science. An orbiting body in space is attracted to another body of greater mass, so that it falls toward the body of greater mass with ever increasing speed as it approaches it. But it never falls into the other body because, by the time it nears it, it is going so fast that it shoots past that body, swings around, and shoots back off into space until it loses its velocity and then, once again, begins to fall back toward the body of greater mass. In other words, it is caught in an orbit around the larger body—just as we now understand that Halley's comet travels continuously around the sun.

But the angle of the swing, the width of the arc a body takes around the sun—whether it is an ellipse, a hyperbola, or a parabola—is solely dependent upon the speed with which that body approaches the sun. A ball thrown straight up into the air, for example, has the same motion as an orbiting comet: When it reaches as high as its momentum will carry it, the ball comes to a dead stop, hesitates for an instant, and then begins to fall back toward the ground, constantly accelerating until it either is caught or hits the ground.

The sun has the same action on a comet as the thrower has on a ball. The sun ''throws'' a comet out into the universe and, when the comet has gone as far as its momentum will carry it, the sun's gravitational attraction pulls the comet back toward it and the comet ''falls'' faster and faster as it returns toward the sun:

Suppose that a small body is at a very great distance from the Sun, and both bodies are motionless. The body will begin to fall toward the Sun, its path being a straight line directed towards the Sun's centre. Another small body, likewise at a distance practically infinite, has a slight motion of its own, but is not moving directly toward the Sun; urged on by the Sun's imperious attraction, its velocity will continually increase: however, as it is not going *directly toward* the Sun, it will not strike it, but as it goes past, the pull of the Sun will cause its path to be violently curved; whirling around the Sun, it will return toward the infinite depths of space from which it came, its orbit becoming a *parabola*. A body which has originally a very considerable velocity of its own will come down to the Sun in an hyperbolic orbit, and then retreat, never again to visit us. A body moving in a parabola may have its velocity checked, as it approaches the Sun, by the attraction of some planet; its orbit will thus be changed to an

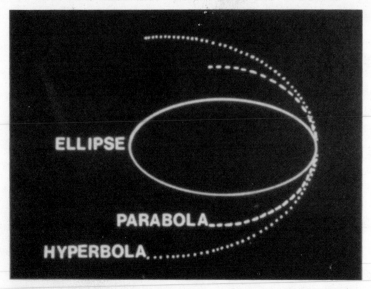

ELLIPSE

PARABOLA

HYPERBOLA

Because comets can only be observed when they are close to the sun, it is difficult to tell whether their paths are ellipses, hyperbolas, or parabolas.

ellipse. Were the movement of the body accelerated by the planet's action, the orbit would become an hyperbola.[5]

In short, if a comet is moving at a speed slow enough to result in a closed curve—an ellipse—it will return again and again to the sun. If it is moving with so great a speed that it swings in a hyperbolic arc around the sun, it will be flung off into space, never to be seen again.

All this, however, was yet to be discovered. Newton's best guess was that comets follow parabolic orbits because that shape would result in the phenomena that had been observed—a very sudden, unexpected approach from space, a close passing-by of the sun (where it would then become visible to us), and then a quick departure back into space.

He knew theoretically that comets could follow ellipses such as the planets took, only with a much more eccentric arc. But he wasn't sure the facts would support this theory. "I am out in my judgement," Newton said, "if they are not planets of a sort, revolving in orbits that return into themselves with a continual motion." And he asked that other scientists verify by observation his assertion that comets did indeed orbit the sun.

Newton's theories concerning comets had not been proven by observation, and that's what Halley planned to do. In so doing, he was to become the first astronomer to apply the law of universal gravitation to a specific problem.

Halley was convinced from the start that comets did, indeed, follow elliptical paths and, therefore, returned to the sun at regular intervals. There was a serious difficulty, though: Since one comet can look remarkably like another (with long or short, straight or curving tails, a brilliant head or one barely visible) and the same comet can change its appearance totally at each return, comets could not be identified by their physical characteristics.

Therefore, if Halley were going to find evidence to support his theory that comets return on a regular schedule to our Solar System, he could not rely on observation and description of comets in history to verify a comet's reappearance. The only way was to plot the path of a comet accurately. In those computerless days, that was no easy task. To understand how difficult it was then, let's look at some of the problems we are aware of today.

We now understand that several physical factors complicate a comet's orbit. Even with our modern computers, plotting the actual path a comet will take can be arduous work.

First of all, from Earth we can observe a comet only when it approaches close to our sun and swings around it. Because the difference between the arcs of an ellipse, a parabola, and a hyperbola is slightest when they are closest to the sun, it becomes very difficult to tell one type of orbit from the other. With such a small part of the comet's path to show what its whole shape is, the tiniest error in calculation or observation can give enormously varied results. Consequently, a comet predicted to return in 100 years might not show up for 25,000.

Then, too, comets are the oddballs of the Solar System. While all the planets, moons, and asteroids revolve in almost circular ellipses around the sun in planes within 13 degrees of one another, the comets revolve mostly in enormous, stretched-out ellipses (often taking more than a million years per orbit), cut across the orbits of the other celestial bodies in the Solar System, and vary in speed from a complete stop to faster than the fastest planet, reaching velocities of 200,000 miles per hour.

Besides this, comets swing around the sun from either direction, whereas the planets all orbit the sun in one direction—counterclockwise. Halley's comet itself orbits in what is known as a retrograde—or clockwise—motion.

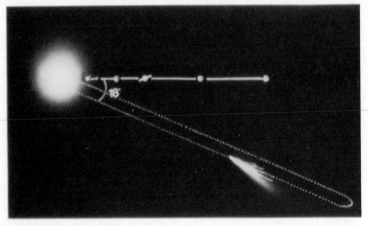

All the planets except Mercury and Pluto orbit the sun on about the same planes. Halley's Comet comes in at an angle of 18 degrees, so far below the orbit of the planets there is little chance of a collision.

One interesting note, however, is that the shorter its period of return, the less likely a comet is to have a retrograde motion. There are, in fact, no comets with a period of less than 30 years that are known to have retrograde motion. This implies that the closer a celestial body is to the sun the more it is influenced by the Solar System and the more it conforms to the laws that govern the other members of the system. (Pluto, for example, the most distant planet, has the most radical orbital inclination—some 13 degrees, whereas the other planets' are all at about 4 degrees. Pluto in fact cuts across the orbit of Neptune—the only planet that invades another planet's orbit; and some astronomers suggest that Pluto is, in fact, not even a true planet.)

Perhaps the greatest difficulty in calculating a comet's orbit, however, is the fact that a comet never follows the same orbit twice.

We now know that comets revolve around the sun in constant elliptical orbits fully regulated by gravity. Therefore, they should return at regular intervals, easily predictable by scientific measurement. Except that they don't.

If the earth orbits the sun every 365¼ days, then why doesn't Halley's Comet orbit the sun every 76 years 22 days, as it did on its last trip in 1910? Why don't comets have fixed orbital periods?

The answer is in the perturbation effect. Massive in size (many are larger than our sun) comets are nevertheless minuscule in mass, so, as they career through space, whenever they pass more massive bodies (such as planets) the gravitational influence of the weightier body can literally pull them out of their orbits. If a comet comes in close to a planet, it may be hurled off into space or, at the very least, have its orbit "perturbed." This can cause it to return to the sun many years, or even centuries, later than expected.

That is why Halley's Comet, for example, commonly thought to have an orbital period of 76 years, as mentioned above, shows up at intervals of anywhere from 74 years to 79 years. Many comets with shorter periods are even more likely to be thrown off course by a planet's gravitational pull because their smaller orbits keep them well within the confines of the Solar System. For this reason they are many times more likely than a long-period comet to come into close proximity to a planet. These comets, bounced around often enough or with enough force, may eventually be thrown out of our Solar System entirely. Jupiter, easily the largest of the planets in our system, is particularly effective as a comet disturber.

But we now know that Jupiter is influential for other reasons, too. The farther from the sun a comet travels, the slower it moves. Therefore, when a comet approaches far-off Jupiter that planet will have a stronger perturbing effect on it than, say, Earth or Mars, because not only does Jupiter have a more powerful gravitational field of attraction but also the path of a slower moving comet can be changed more easily. That is why Jupiter has played a pre-eminent role in the history of comets. In fact, for centuries some scientists believed that comets were born from the planet Jupiter and spat out into their present orbits by a mammoth explosion on the face of the planet. (More about these theories later.)

One of the more interesting cases of a planet's perturbing effect involved Lexell's Comet. Charles Messier, the famous French astronomer, in 1770 discovered a comet which at first was unimpressive. It looked to Messier like nothing more than a faint nebula. But as the months went by it grew to immense proportions, dominating the sky until its head was five times the diameter of the full moon. It remained visible for six months.

Meanwhile, in St. Petersburg, the astronomer Anders Johann Lexell computed its orbit to be 5 years, 7 months. This was the first time this particular comet had ever been seen. And it was to be the last.

Everyone was puzzled by Lexell's Comet's disappearance until in the next century, after a long, laborious investigation, two scientists named J.C. Burckhardt and Urbain Leverrier concluded that, previous to Messier's discovery, Lexell's Comet had taken a totally different orbit. In 1767, three years before Messier spotted it, Lexell's had made such a close fly-by of Jupiter that, they calculated, the planet had exerted enough gravitational attraction on the comet—three times that of the sun—to force it into a new orbit. Whereupon Messier had spotted it. On its next go-round, however, in 1776, they figured it had been in a poor viewing orientation and no one had stumbled on it. In 1779, on its next return to the sun, it again made a close encounter with Jupiter, and this time it was so violently perturbed by the giant planet that it was thrown out of its elliptical orbit into a hyperbolic one. It has never been seen since.

The perturbing effect of the planets on comets is not always as pronounced as that upon Lexell's Comet. Most comets undergo frequent, but smaller, orbital deviations on each circuit.

Encke's Comet is a more typical example. It was discovered by yet another great French comet hunter, Jean Louis Pons, who first spotted it at Marseilles on November 26, 1818. It was a small but very bright object and remained visible about seven weeks. Its orbit was calculated by Johann Franz Encke at about 3 years 6 months. Encke then did some backchecking and found that comets with similar orbits had been noted in 1786, 1795, and 1805. He felt sure it was the same object that, between 1786 and 1818, must have passed the sun seven times without being noticed. On its next return—in 1822—Encke's Comet was nine days late, and on its subsequent returns it has shown similar small but definite orbital deviations.

Edmund Halley knew nothing about all this, of course, but he knew the right questions to ask. As early as 1680 he had tried to plot the path of a comet according to straight lines and, of course, had failed. But it was then that he first got the notion, borrowed from Cassini, that comets did, in fact, return again and again to the neighborhood of our sun:

> Monsieur Cassini did me the favour to give me his books of Comett Just as I was goeing out of towne; he, besides the Observations thereof, wch. he made till the 18 of March new stile, has given a theory of its Motion wch. is, that this Comet was the same with that appeared to Tycho Brahe 1577, that it performes its revolution in a great Circle including the earth wch. he will have to be fixt in about two yeares and halfe.[6]

Somewhere around 1695, Halley once again set out in earnest to plot the orbit of a comet. This time he meant to prove what many suspected and Newton had asked to have proven—that comets do return to the sun. Armed with Newton's new theories of universal gravitation, he knew he could successfully track one.

As we already know, he reckoned that if he could find two or more comets that followed the same orbit years apart, he felt confident he could prove they were the same comet. For instance, he knew that the comets of 1680 and of 1682 had very different orbits, and so they were not the same comet.

His first step was to chart the paths of previously observed comets from whatever reports he could gather. Halley, with his customary thoroughness, investigated every known sighting right back to the ancient Greeks. But, because until nearly the 16th century comets were generally thought to be simply upper atmospheric disturbances, little of value was available. The comets of 1337 and 1472 were the only two for which Halley found accurate enough data to be useful. That is, their discoverers had reported both time and place of observation and the position of the comet among the fixed stars.

Fortunately, the 16th and 17th centuries were full of good sightings. Also, between 1680 and 1695 (when he was doing the final mathematical work on comet periodicity) Halley himself observed five additional comets. In all, he collected data on the orbits of 24 comets.

The comet of most interest to Halley, however, was the one that had been seen in 1682. Halley carefully checked his list of recorded comets and found that the pathway of 1682 paralleled the orbit of a comet seen in 1607 (observed by Kepler) and another in 1531 (seen by Apian)—about 76 years apart. These observations were confirmed

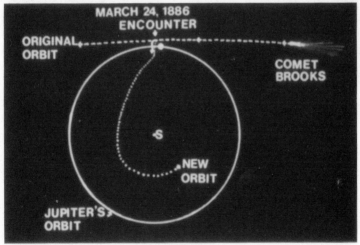

The path of Brooks's Comet was perturbed, by its extremely close passage of Jupiter, into a much shorter orbit.

Artist's rendering of Brooks's close passage of Jupiter. The comet actually passed between Jupiter and its moon Io.

by comet sightings of 1456, 1380, and 1305—all about 75 years apart. Could these three comets, Halley asked himself, be one and the same?

He set out to prove it.

Halley used Flamsteed's original observations and then recalculated the entire orbit for accuracy. Of course he had to take into account the perturbations on the comet's orbit by Jupiter and Saturn, which meant he had to calculate those planets' positions, as well. Then he had to adjust the flight of the comet to compensate for those encounters and recalculate the orbits. These results had to be combined with the position of Earth at the time when the comet

(U.S. Naval Observatory Photograph.)

Halley's Comet in 1910. The computer-enhanced colors of this plate show vividly the various levels of brightness difficult to see in an ordinary photograph. © 1978 AURA, Inc., National Optical Astronomy Observatories, KPNO.

Comet Ikeya-Seki, November 6, 1965. (*Jet Propulsion Laboratory*)

The Bayeux Tapestry commissioned by William the Conqueror portrays Halley's Comet of 1066 as the precursor of the Norman Conquest of England. King Harold looks on in alarm.

Section of Bayeux Tapestry showing close-up of Halley's Comet.

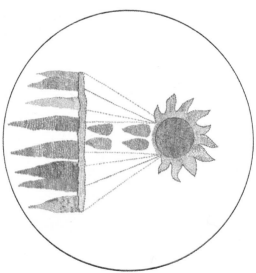

The magnificent tail displayed by the Comet West. (*Photo by Peter Stättmeyer*)

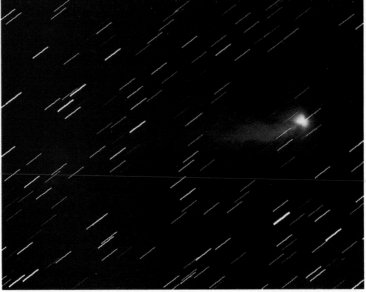

Comet Humason, an exceptional comet, developed its tail while still more than five times as far from the sun as the earth is. It was an impressive telescopic object while still as distant as Jupiter. It has a period of 2900 years. © *California Institute of Technology.*

The Great Leonid Meteor Shower (photographed in 1966) is created by the orbiting debris of a comet. (*Jet Propulsion Laboratory*)

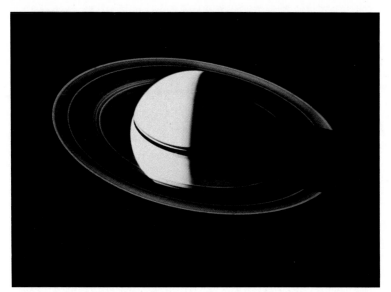

The perturbing influence of Saturn delayed the 1758 appearance of Halley's Comet by 100 days. (*Jet Propulsion Laboratory*)

According to the "Ejection Theory," comets were spit out of the giant planet Jupiter. (*Jet Propulsion Laboratory*)

Comet Ikeya-Seki. Aristotle believed that comets were merely disturbances of the upper atmosphere.
Photo by Charles F. Capen.

neared the sun and where it would be in relation to the stars in our sky. All in all, an exhausting piece of work.

It was in June 1705 that Halley published his results in his pioneer paper, *Astronomiæ Cometicæ Synopsis*. In it Halley surveyed comet sightings from ancient times to Newton, and, for the future use of astronomers, he included a General Table "to endure as long as astronomy" by which the flights of other comets could be calculated. Then he made his famous prediction:

> Now many things lead me to believe that the Comet of the year 1531, observed by Apian, is the same which, in the year 1607, was described by Kepler and Longomontanus, and which I saw and observed myself, at its return in 1682. All the elements agree, except that there is an inequality in the times of revolution; but this is not so great that it cannot be attributed to physical causes. For example, the motion of Saturn is so disturbed by the other planets, and especially by Jupiter, that his periodic time is uncertain, the extent of several days. How much more liable to such perturbations is a comet which recedes to distance nearly four times greater than Saturn, and a slight increase in whose velocity could change its orbit from an ellipse to parabola? The identity of these comets is confirmed by the fact that in the summer of the year 1456 a comet was seen, which passed in retrograde direction between the Earth and the Sun, in nearly the same manner; and although it was not observed astronomically, yet, from its period and path, I infer that it was the same comet as that of the years 1531, 1607, and 1682. I may, therefore, with confidence predict its return in the year 1758. If this prediction be fulfilled, there is no reason to doubt that the other comets will return.[7]

Halley then appealed to astronomers of the next generation to watch carefully for his comet, adding in a humble tone, knowing full well he would not be around to see whether his prediction panned out:

> Wherefore, if it should return according to our prediction about the year 1758, impartial posterity will not refuse to acknowledge that this was first discovered by an Englishman.

This was surely the most courageous and, at the time, outrageous, prediction ever put forth in scientific history. Newton had made it all feasible with his gravitational laws; but Newton's theories were brand new and as yet untested. To suggest that a comet, certainly the most mysterious, fear-provoking of heavenly bodies, was as predict-able as the sunrise was putting one's neck way, way out.

To many, in fact, Halley's prediction seemed an outlandish bit of grandstanding. In the first place, because Halley probably wouldn't be around to see whether his prediction came true (he would be 102 at its next return), he wasn't really risking much—he was getting lots of attention without really putting his reputation on the line. Besides, the prestigious position of Royal Astronomer was then coming vacant, and many believed this theorizing was all just a public-relations move to solidify Halley's chances at the job.

All this skepticism is understandable if you remember that for 2000 years comets had been thought of as sinister and wildly unpredictable. Superstition, despite the Age of Enlightenment, was still the most powerful governing force of the common man—and of quite a few men of science, for that matter: Astrology, not Astron-omy, was still the most influential science of the heavens. To state without doubt that those erratic, mysterious balls of light followed the regular movements of the planets was a bold presumption.

As 1758 approached, there was a great stir throughout Europe. One contemporary historian notes that the astronomers of Europe did not go to bed that year for fear of missing the long-expected comet. Who-ever found it first could build his reputation on the discovery. But no one knew for sure when it would show up because of the enormous fluctuations in the comet's period. The 1607 return had taken 27,811 days; the 1682 return, 27,325 days. A difference of 459 days—nearly 15 months. The search was on!

Fortunately, mathematical techniques had taken a leap forward since Halley's prediction some fifty years before, and as the comet's return drew near, Alexis Claude Clairaut, the French mathematician, undertook the enormous task of calculating the exact date of its peri-helion—the moment it would come nearest to our sun. That meant he had to figure not only the comet's changing speeds but also the effect of all the other celestial bodies it would encounter, principally Jupiter and Saturn. So he needed the help of an astronomer, and he enlisted Joseph Jérôme de Lalande, the famous French astronomer. These two were also joined by a woman with no scientific background at all, the wife of Jean André Lepaute, clockmaker to the French king, Louis XVI.

The reasons Clairaut picked her were a mystery to everyone but Clairaut, for she was singularly unqualified for the job. Yet without her, the task would surely have failed. She had the energy to make it all work. She was able to carry on when the rest of them had driven themselves nearly to death. Lalande, in fact, suffered an ailment while tackling the calculations from which he said he "never recovered."

The task was arduous:

> Having devised a method which appeared to possess all needful accuracy, he [Clairaut] commenced in conjunction with the celebrated Lalande and a lady, Madame Lepaute, the immense mass of calculations requisite for the complete attainment of his object. It was necessary to compute the distances of the comet from the disturbing planets, Jupiter and Saturn, not only from 1682, when it was last observed, but for the previous revolution, or for a space of more than 150 years. This of itself was a most laborious business; but the succeeding part of the work, where the disturbing force of each planet was required for this long period, involved much greater and more intricate calculations. Lalande minutely describes the plan adopted: for six months they computed from morning to night.[8]

Having started only six months prior to the date when Halley had predicted his comet would show up (Christmas 1758), their time was running short. The three worked day and night, often forgoing food and sleep as the comet came ever closer.

Finally, on November 14, 1758, Clairaut submitted his findings to the Academy of Science. His calculations showed that the perturbing influence of Saturn would delay the comet by 100 days. Its close contact with Jupiter would slow it down another 518 days. He further added that Halley's Comet traveled so deeply outside our own Solar System that it might be influenced by some planet or some other body not yet even known. (As the planets Neptune and Uranus had not yet been discovered, Clairaut's statement was insightful, indeed.) Therefore, Clairaut predicted, the comet would be delayed by some twenty months, with a possible error of a month either way because of "unknown influences." He predicted perihelion (the point at which the comet passes closest to the sun) for April 13, 1759. If he was right, it meant that the first sighting of the comet, just then (in November 1758) nearing the region of Mars, should occur almost any day.

Despite the fact that several professional astronomers, including Messier, now undertook 24-hour watches, it was a Saxon farmer and amateur astronomer in Prohlis, near the city of Dresden, who first spotted the return of Halley's Comet on Christmas Day, 1758, exactly as Halley had predicted. Johann George Palitzsch, who loved nothing better than to scan the sky for hours, stood watch most of that very cold Christmas night with only an 8-foot telescope. In fact, much to the chagrin of the professionals, Palitzsch was said to have first spotted the comet with his naked eye!

The first professional astronomer to discover the comet was Charles Messier, in France, on January 21, 1759. But instead of immediately disclosing the observation to the scientific community, his superior, Delisle, director of the Paris Observatory, ordered him to keep it a secret. Some 80 years later, J.R. Hind described it this way:

> Delisle would not allow him [Messier] to give notice to the astronomers of that city [Paris] that the long-expected body was in sight, and Messier remained the only observer before the comet was lost in the sun's rays. Such a discreditable and selfish concealment of an interesting discovery is not likely to sully again the annals of astronomy. Some members of the French Academy looked upon Messier's observations, when published, as forgeries, but his name stood too high for such imputations to last long, and the positions were soon received as authentic, and have been of great service in correcting the orbit of the comet at this (1835) return.[9]

Halley's Comet reached perihelion on March 12, 1759. It was seen all over Europe throughout that spring. Sir Edmund Halley was lauded in all circles for his great prediction, and the Great Comet of 1682, which had originally caught his attention, was ever after known as "Halley's Comet."

CHAPTER 6

THE LEGACY OF SIR EDMUND HALLEY

The arrival of Halley's Comet as predicted put a solid stamp of credibility not only on the "New Astronomy" but also on all of science. It was becoming easier for people to accept the fact that the heavens (and the earth, for that matter) were not an unfathomable mystery. On the contrary, the universe could now be shown to be clearly governed—and made predictable—by laws of nature.

Such a triumph produced a deep impression in the scientific world. As Lalande enthusiastically put it:

> The universe beholds this year the most satisfactory phenomenon ever presented to us by astronomy; an event which, unique until this day, changes our doubts to certainty, and our hypotheses to demonstration. . . . M. Clairaut asked one month's grace for the theory; the month's grace was just sufficient, and the comet has appeared after a period of 586 days longer than the previous time of revolution, and thirty-two days before the time fixed; but what are thirty-two days to an interval of more than 150 years, during only one two-hundredth part of which observations were made, the comet being out of sight all the rest of the time! What are thirty-two days for all the other attractions of the solar system which have not been included; for all the comets, the situations and masses of which are unknown to us; for the resistance of the ethereal medium, which we are unable even to estimate, and for all those quantities which of necessity have been neglected in the approximations of the calculation?[10]

Despite Lalande's gushing tribute, Halley was not fully appreciated in his own day for his truly great accomplishments in astronomy. That may have been largely Halley's own fault, for his interests, knowledge, and expertise extended to so many fields and endeavors that the sheer weight of his other accomplishments at the time made his comet prediction pale in comparison. The 19th century mathematician Augustus De Morgan described the breadth of Halley's knowledge:

> There is no one of the multifarious branches of knowledge which Halley cultivated in which he did not prove himself capable of surpassing all his contemporaries, except only Newton in mathematics and physics. Such varied knowledge, so deep in all its parts, such universal energy, so equally distributed through a long life, have hardly a parallel. If any one were to ask which we thought most likely, another Halley or another Newton, that is as extraordinary a man as the former or as the latter, we should reply—without denying the vast superiority of Newton in those points in which he was superior—that we should think the second more reasonably to be expected than the first. Wherever Halley laid his hand, to do work cut out by himself, he left the mark of the most vigorous intellect, the soundest judgment, the most indomitable courage against difficulties.[11]

The depth and range of Halley's accomplishments boggles the modern mind in this age of specialization. His interests knew no limits.

Halley made significant and lasting contributions to astronomy, physics, mathematics, geophysics, navigation, and archeology. A sailor himself, he found a way for sailors to determine longitude at sea, developed one of the first diving bells, was the first to chart the wind patterns of the world, and successfully linked the position of the moon with the earth's tides. At the same time, he was a pioneer in population statistics, and his tables of mortality and age served as the basis for life insurance calculations. In addition, by applying scientific investigation to classical studies, he figured out the exact time and place where Julius Caesar first invaded Britain in 55 B.C.

Besides his scientific accomplishments, Halley also was appointed a captain in Britain's Royal Navy. He clearly loved adventure, and to that end he made two heroic sea voyages as captain of the *Paramour,* a man-o'-war, upon which he survived mutiny, typhoid, and the icebergs of the Antarctic.

This excerpt from the captain's log of his first voyage shows that Halley did not live the life of Academe, holed up in some ivy-covered tower:

Between 11 and 12 this day we were in iminant danger of loosing our Shipp among the Ice, for the fogg was all the morning so thick, that we could not see for long about us, where on a Sudden a Mountain of Ice began to appear out of the Fogg, about 3 points on our Lee bow: this we made a shift to weather when another appeared more on head with severall pieces of Ice round it; this obliged us to tack and had we mist Stayes (putting on another tack), we had most certainly been a Shore on it, and had not beene halfe a quarter of an hour under way when another mountain of Ice began to appear on our Lee bow; which obliged us to tack again with the like danger of being on Shore: but the sea being smooth and the Gale Fresh wee got Clear: God be praised. This danger made my men reflect of the hazards wee run, in being alone without a Consort, and of the inevitable loss of us all in case we Staved our Shipp which might soe easily happen amongst these mountains of Ice in the Fogg, which are so thick and frequent there.[12]

Besides his sea adventures, Halley also served two missions as a diplomat and, after a great coinage scandal in Britain, he was named by Newton as Comptroller of the Royal Mint to help stop widespread embezzlement.

Halley is most famous, of course, for his work in astronomy and, in that field, he had no equal.

He was the first to establish that the stars are not fixed but have their own movement—a revolutionary idea in that era. Halley realized that, although the stars were reported in the same positions from the earliest recorded history, their vast distances from the earth could conceal even large real motion from the naked eye. Only precise telescopic observations, made years apart, could reveal this movement. Of course he was right; but it would be well over a century before another observer—the Italian astronomer Giuseppe Piazzi—would follow up on his ideas.

Halley also was the first to establish the sun's true distance from Earth and to study solar eclipses closely. His work on the moon's parallax and the moon's motion were not fully understood for generations. And he was centuries ahead in his assertion, generally believed by astronomers today, that the universe has no center.

With the death of John Flamsteed on December 31, 1719, the post of Astronomer Royal became vacant. In February of 1720, King George I commissioned Halley as the second person to hold that title.

Halley was then 64 years old, but still a very energetic and enthusiastic man. He promptly turned the badly equipped Greenwich

Observatory into one of the finest in the world. His wife, Mary, his close companion for 55 contented years, died January 30, 1736. The next year Halley suffered a minor stroke, which caused some paralysis; but although he now had to have an assistant, he worked on.

In 1741, Halley's paralysis deepened ". . . and thereby his strength wearing, though gently, yet continually, away, he come at length to be wholly supported by cordials as were ordered by his Physician." Finally, on the night of January 14, 1742, "being tired . . . he asked for a glass of wine, and having drank it presently expired as he sat in his chair without a groan . . . in the eighty-sixth year of his life."[13]

At Halley's request, he was buried next to his wife in the small churchyard of St. Margaret's in Lee. It is not far from Greenwich, where he had done most of his work. His daughters placed the following epitaph on their father's tomb:

Under this marble peacefully rests, with his beloved wife, Edmundus Halleius, LL.D., unquestionably the greatest astronomer of his age. But to conceive an adequate knowledge of the excellencies of this great man, the reader must have recourse to his writings; in which almost all the sciences are in the most beautiful and perspicacious manner illustrated and improved. As when living, he was so highly esteemed by his countrymen, gratitude requires that his memory should be respected by posterity. To the memory of the best of parents their affectionate daughters have erected this monument in the year 1742.

PART THREE
ALL ABOUT COMETS

CHAPTER 1

COMET FAMILIES

Scientists have estimated that some five million comets belong to our Solar System alone. In fact, comets are the most numerous objects in the universe larger than meteorites—and most of those are smaller than a grain of sand.

To date, scientists have accurately plotted the orbital paths of more than 600 of them. These comets are usually divided into four main classes, according to the amount of time it takes each one to orbit the earth. They are: short-period, medium-period, long-period, and nonperiodic comets.

Short-period comets are regular visitors to our sun and have well-known orbits. The range of their periods is from three to twenty years, with most returning every seven to eight years. Encke's Comet, with a period of 3½ years, has the shortest period of all known comets. Griss-Skjellerup is next at 5.1 years. It has now been witnessed on more than fifty different orbits since it was first discovered in 1786.

Most short-period comets are very faint and do not display a tail. Therefore, although they travel only a relatively short distance from the sun, they still cannot be tracked throughout their entire orbit.

Most of the more than a hundred known short-period comets belong to the Jupiter Family. That is, their aphelia (the points where each is farthest from the sun) are near the orbit of Jupiter. This implies that these comets came under the strong gravitational influence of Jupiter at one time and were ''captured.''

Scientists formerly believed that there were many planets with their own comet families, such as Neptune and Saturn. Halley's

Comet, for example, was said to be a member of the Neptune Family. But this theory has since been disproven. There is little doubt that the Jupiter Family is real, however. The average aphelion of these comets is between 5 and 6 Astronomical Units (the Astronomical Unit is approximately the distance between the sun and Earth). The orbit of Jupiter is 5.2 A.U.

Medium-period comets have orbits larger and more elongated than those of short-period comets. Crommelin's Comet and the Comet Schwassmann-Wachmann I are typical medium-period comets, with aphelia that range as far as the orbit of the planet Uranus. Their periods range from around 20 years to 60 years.

Long-period comets have orbits of from 60 years to several thousand years. Halley's Comet, with a period of about 76 years, is included in this class. Donati's Comet, one of the brightest and most famous of all comets, which last appeared in the 19th century, has a period of 2000 years.

Nonperiodic comets are those whose orbits are so large or so eccentric that they can't be predicted. In fact, most of the comets in our Solar System have never been seen and probably never will be. They simply travel distances that are beyond the range our present telescopes can detect. A.C.D. Crommelin has estimated that most of these comets have periods that average about 40,000 years and that about three hundred come to perihelion—make their closest approach to our sun—each year.

The longest computed orbit, figured in this century by George van Bjesbroeck, was that of the Comet Delavan of 1914, estimated to have a distance at aphelion of 15,810-*billion* miles and a period of 24-*million* years!

Most of the really bright comets are in this so-called "nonperiodic" category. For example, the Great Comet of 1864 is estimated to return every 2.8-million years, and Comet Kohoutek, a recent visitor in 1973, is thought to have a period of 75,000 years.

As we have seen, there are several million comets orbiting our sun and several thousand have come within view of our Earth since Halley's Comet last flew by us. Also, Halley's is not nearly the biggest or the brightest comet ever seen, although it is easy to see with the naked eye. Then why do human beings anticipate with such excitement the arrival of this celestial object every 76 years? Why is Halley's the most famous comet?

Simply put, Halley's Comet is popular because it is the only bright comet with a period of less than a century. All other brilliant comets have periods of hundreds, or thousands, or even millions of years. Halley's orbital timespan and the lifespan of a human being are both, conveniently, about 75 years; so it is the only conspicuous comet—easily observed without a telescope—that just about everyone will get to see in a normal lifetime.

Halley's large size, its full—often spectacular—tail, and its perihelion about midway between the sun and the earth all make for a good naked-eye view of the comet. It's something you can tell your grandkids about, in other words. And if they're lucky, someday they can see it with their own eyes.

Comet West had a period of 16,000 years, but on its last orbit may have been perturbed out of the Solar System. *Lick Observatory photograph.*

CHAPTER 2

HEADS AND TAILS:
THE STRUCTURE OF A COMET

A bright comet is made up of three distinct parts—the nucleus, the coma, and the tail.

The *nucleus* is the distinct starlike core of the comet's head. It is composed of the frozen gasses and dust from which the other parts of the comet (the coma and the tail) draw their material. Usually a comet has one nucleus, but sometimes comets are seen with two or more nuclei.

No photograph of a nucleus exists, and even the largest telescopes have not been able to reveal its mysterious makeup. Although some researchers have said that they observed tiny disk-shaped planets within the nucleus, these claims have never been confirmed.

The general diameter of comets ranges from a few hundred yards to about five miles. The Comet Pons-Winnecke made a particularly close approach to the earth—within three and a half million miles—in 1927. A French astronomer named Baldet, upon studying this average-sized comet with a 30-inch telescope, concluded that the comet's nucleus was no more than a half-mile in diameter.

The mass of a typical comet is estimated at somewhere between 100-million metric tons and 10-trillion metric tons—and most of that mass is contained in its tiny nucleus.

The *coma* surrounds the nucleus. It is a spherical cloudy mass of gasses, frozen gasses, and dusty particles of materials such as iron, nickel, and magnesium. The nucleus and the coma, taken together, are usually referred to as the ''head'' of the comet.

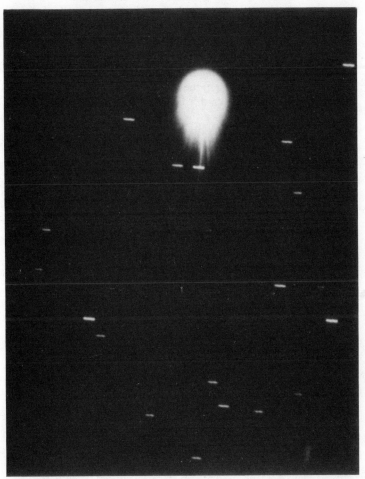

The head of Finsler's Comet on August 28, 1937. *Lick Observatory photograph.*

In some cases the coma can be huge: It can extend as far as a hundred thousand to a million miles from the nucleus which feeds it. The Great Comet of 1811 had a diameter of 1.25-million miles—larger than the sun.

The tail is a sometime thing. The spectacular display offered by a comet when its head is enlarged and its tail long and streaming is quite unlike its normal appearance. In fact, for most of its journey around the sun it is simply an unspectacular ball of ice—a kind of astronomical wallflower. It is only when it draws nearer to the sun that the comet grows in size and brightness.

When the comet is beyond the orbit of Jupiter in its "undeveloped" state, it simply is too small and too far away to be seen by even our strongest telescopes. But at about three A.U. (three times the earth's distance from the sun) the comet begins to develop its cloudy coma. At about twice the earth's distance from the sun, the tail starts to form—and the comet has grown in size from as little as 200 yards across to the largest single object in the universe, stretching out sometimes 25-million to 30-million miles.

As the comet passes behind the sun it becomes lost to our view in the sun's rays, and "disappears." But it soon reappears, often more brilliant and with an even longer tail.

As it flies back off into the distant parts of the Solar System, the

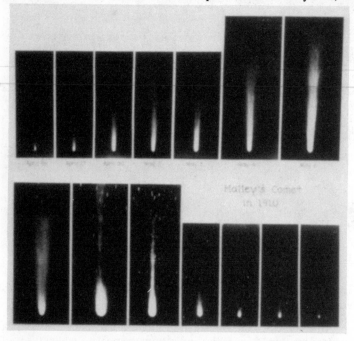

14 views of Halley's Comet taken between April 26 and June 11, 1910, show the dramatic development and disintegration of the tail of a comet as it approaches the sun and then, once again, journeys back into the far reaches of the Solar System. Mt. Wilson Observatory.

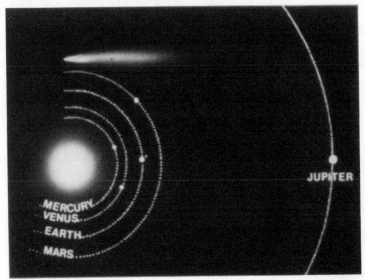

The Great Comet of 1843 reportedly had a tail that stretched 200 million miles—greater than the distance between the earth and Mars.

tail again gradually diminishes, then the coma grows dim and shrinks. Eventually, only the nucleus is left to make the lonely journey into the galaxy.

It is the tail that determines whether a comet will make a spectacular display. A relatively small comet may have a long, bright tail, whereas, a larger comet—one with a nucleus hundreds of miles in diameter—may never grow a tail at all and remain, even when seen through the most powerful telescope, just an indistinct blur.

There are, in fact, many more comets without tails than with them. Because we almost never see the tail-less comets, they pass unnoticed, while a comet with huge plumage grabs everyone's attention. The Great Comet of 1843 reportedly had a tail more than 200-million miles long, greater than the distance between the sun and Mars. It extended halfway across the night sky.

Then, too, many comets have tails that are just too faint to be detected, or whose light inhabits a wavelength beyond the capabilities of our eyes to see, but which are frequently revealed by blue-light-sensitive photographic plates.

Ultimately, though, whether or not a tail forms depends principally on how close the comet approaches the sun. The closer it gets, the more material will ordinarily be evaporated from the nucleus of the comet. It took a long time for scientists to understand this.

One of the earliest speculations about the nature of a comet's tail was put forth by the philosopher Panaetius, who didn't believe the tail existed at all. He thought it was an illusion created by the sun's rays. His ideas held sway well into the 17th century, when, in the year 1610, Kepler came forward with a remarkably modern view of tail formation, as follows:

> Gross matter collects under a spherical form; it receives and reflects the light of the sun and is set in motion like a star. The direct rays of the sun strike upon it, penetrate its substance, draw away with them a portion of this matter, and issue thence to form the track of light we call the tail of the comet. This action of the solar rays attenuates the particles which compose the body of the comet. It drives them away; it dissipates them. In this manner the comet is consumed by breathing out, so to speak, its own tail.[14]

Next, Isaac Newton reviewed observations of comets throughout history and concluded that the tail was a product of "the atmosphere" of the comet itself. "A very small amount of vapor," he suggested, "may be sufficient to explain all the phenomena of the tails of comets."

Basically, Newton had it right again.

It was not until the 19th century, however, when F.W. Bessel and Wilhelm Olbers first postulated the theory that electrical forces may cause a comet's tail to form, that any real scientific answer was presented.

Early speculation aside, the key to understanding what causes the formation of a comet's tail is the curious fact that, no matter what part of the sky the comet is in or which direction it is coming from, the tail, no matter how large or small, *always points away from the sun*. This means that after it has passed perihelion and rounded the sun the tail *precedes* the head. Instead of following, like the smoke from a fast-moving train, the tail now leads the way.

So, whether a comet is high or low in the night sky, whether it is in the eastern or western or southern or northern sky, its tail will point away from where the sun has set. Also, if the comet is seen either in the western sky at sunset or the eastern sky before dawn, the tail will appear to point upward.

Ancient Chinese sky watchers noted as early as A.D. 837 that a comet's tail always pointed away from the sun. Although in second-century Rome Seneca had remarked that "the tails of comets fly from

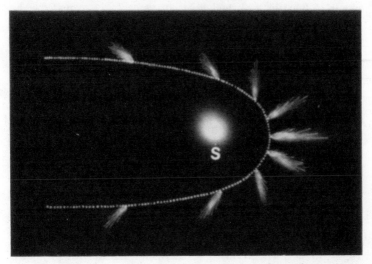

The tail of a comet always points away from the sun.

the sun's rays,'' it is generally accepted that Peter Apian, in 1540, was the first European astronomer to write about this peculiarity in his book *Astronomicum Cæsareum*. But Tycho Brahe, the greatest astronomer of that time, refused to believe it because he thought Apian had not proved his point sufficiently.

Because the tail pointed away from the sun, it was reasonable to assume that the sun had something to do with this behavior. The sun's gravity, of course, exerts a great *pull* on the comet. But there had to be a force in the sun that *pushes* the matter in the tail away.

Robert Hooke, Newton's old nemesis, was the first to suggest that this repelling, pushing force was the pressure of the sun's light rays. Almost three hundred years later the theory was finally proved in the laboratory when three physicists—Lebedev, Nichols, and Hull, in 1901—floated mushroom spores past a light source. Their instruments recorded that the minute spores were in fact pushed off course by the light.

Lebedev and his colleagues had proved that light does exert a force. This force is extremely small by normal standards: about one 10-millionth of a pound per square foot. But then, a comet's tail is composed of extremely small particles; and solar radiation pressure can move particles as large as an inch in diameter.

Now there was an explanation for the growth of a comet's tail as it approached the sun: When the comet was near its aphelion in the deep freeze of space, its gasses remained frozen, but as it came ever closer to the sun, heat and light correspondingly increased. This caused the gasses of the comet to evaporate, forming the enlarged coma (or head). Then the sun's radiation blew material from the coma into a tail. The explanation seemed to fit the observed data.

For a time the solar-radiation-pressure theory appeared to account for the formation of all comet tails. But this theory was sorely tested in 1908, when Comet Morehouse was seen. The tail of this comet put on a furious display, with great bunches of material being ejected from the head out into the tail at speeds and in volumes that could not be accounted for by simple radiation pressure alone. In fact, the force of the material being ejected from the comet's nucleus was measured by Sir Arthur Eddington, chief of the Royal Observatory in Greenwich, as 800 times the pull of gravity. Radiation pressure could account for a force of about 25 times the pull of gravity, but nothing yet known could explain these magnitudes of repulsion.

It would remain a mystery for some time. It was not until the 1950s, in fact, when Ludwig F. Biermann, the German physicist, demonstrated that the sun was sending out a constant stream of particles in all directions, that the puzzle began to be solved: The tail of a comet acts like a great cosmic windsock, showing clearly which way the so-called "solar wind" is blowing. Without exception, it was shown to be blowing away from the direction of the sun. But what causes the solar wind, what is this constant stream of particles?

The corona—the outer atmosphere of the sun—is a layer of superheated gasses that surrounds the sun the way air surrounds the earth. Now scientists have discovered that the corona actually extends throughout the entire Solar System. And it may even extend as far as 160 A.U., four times the distance from the sun to its farthest planet. For years it was assumed that the corona was a static atmosphere. But it turns out that, the farther the corona is from the sun, the faster it expands because the layer of gasses—and, thus, resistance—lessens. At 6-million miles away from the sun the corona is expanding faster than the speed of sound—it has become a supersonic wind. And it continues to increase its pace until the particles in the corona are traveling at several times the speed of sound:

> The coronal gas streaming slowly away is accelerated very gradually: it takes about five days and about a million kilometers of travel to really get under way. Thereafter it speeds up to hundreds of kilometers per second, and in about nine days it has spanned the 93-million miles to the Earth. The gas we see at the bottom of the corona on a Sunday will be passing us about Tues-

The bright Comet Arend-Roland developed a "spike" tail that was really a fan of large dust particles seen on edge. Top: *Palomar Observatory photograph.* Bottom: *Lick Observatory photograph.*

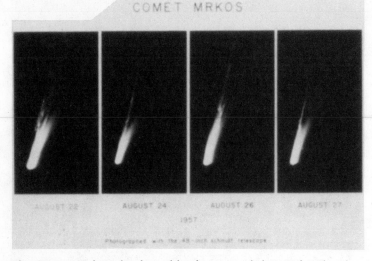

The Comet Mrkos displayed both gas and dust tails. The dust tail is short, broad, and curved; the gas tail is straight, long, and narrow. *Palomar Observatory photograph.*

day of the following week. Two weeks after this gas zooms by us, it will pass Jupiter.[15]

This solar wind of hydrogen blows constantly throughout our Solar System, pushing past the earth at 900,000 miles per hour, sweeping up gasses, meteoric dust, cosmic rays, and the material of comets in its wake. The solar wind, then (and, to a lesser degree, solar radiation), is the major cause of comet tail formation.

The size and shape of each comet's tail is unique. Some are slender streaks; others bend back in a long graceful arc; still others have a sort of squat, arched appearance. It was the Russian astronomer Thomas Bredichin who in 1890 first classified comet tails. He reported three common types: *long, straight rays; curved, plumelike trains;* and *short, stubby, sharply curved brushes of light.* These different types of tails are distinct, but a comet may have more than one type of tail.

The type of tail or tails a comet forms is due entirely to the tail's composition. Tails are basically either gaseous or dusty. In general, gas tails are straight and dust tails are curved.

The way dust tails develop is fairly uncomplicated: As we have seen, heat and radiation pressure sweep dust particles out of the head.

Then, as the comet moves swiftly through the Solar System, the heavier dust particles tend to lag behind, spreading out along the comet's curved orbit and forming a curved tail.

Formation of the gas tails, however, is more complex. They are created by the interaction of the comet with the solar wind. The charged (ionized) particles of the comet's tail are trapped into lines by the magnetic forces set up by the solar wind. This results in a tail that is usually narrow and very straight.

The short-period comets usually have the narrow gas tails, which are composed mostly of ionized carbon monoxide. The longer-period comets, such as Halley's, usually have tails made of the more solid dust particles, and so their tails are more often of the spectacular, sweeping variety.

Many comets have a combination of both gas and dust tails. Donati's Comet, for example, had a magnificent, curved tail of dust material and *two* straight and narrow tails of gas.

CHAPTER 3

WHAT A COMET IS MADE OF

Comets are the biggest objects in our Solar System. Just the head of an average-size comet is larger than Earth—and yet it will weigh only one 10-billionth as much as our planet. Accordingly, comets have been described as "great big bags full of nothing." Only in the last 30 years, however, has anyone come close to explaining exactly what's in this great big bag.

One of the first viable theories about the composition of a comet was R.A. Lyttleton's Flying Gravel Bank model. According to Lyttleton, a British astronomer, the head of the comet is composed of dust particles, forming an apparently solid nucleus near the comet's center. The nucleus, therefore, is not a cohesive mass but rather a formation of independent pieces, all orbiting the sun together in a path that coincides with the comet's path and held closely together by the sun's enormous gravity. As the comet approaches perihelion, the particles crowd even closer together and begin to collide with one another, breaking into still smaller bits. This even finer dust is then driven out of the head by solar radiation and solar wind. As the comet approaches the sun it also warms up, and gasses trapped among these dust particles begin to escape, forming the tail.

The Flying Gravel Bank theory effectively explains many observed phenomena of cometary behavior, such as meteor streams (which will be explained later), comet dust and gas, and the heightened activity of a comet as it nears the sun.

The theory does not explain several other phenomena, however. For example, just how much gas could be held between the dust particles? At every orbit a comet would lose gas as the comet disintegrated. Most bright comets, in fact, lose several tons every second for many months as they enter the inner Solar System. If a comet were a collection of particles "holding" gas between them, one passage by the sun would just about tap out its supply. Yet we know of at least 29 appearances of Halley's Comet, and, by all reports, its brightness seems not to have diminished at all.

Lyttleton has suggested that the comet could replenish its supply of gasses as it moves through its outer orbit. But interplanetary space contains so little matter that this aggregate of dust could never absorb enough gas to make up even a fraction of the material lost on each orbit of the sun.

A second strong argument against the Flying Gravel Bank theory concerns the so-called "sun-grazing comets," such as Comet Ikeya-Seki of 1965. The sun-grazers come so close that the sun's heat would vaporize virtually every common substance; dust particles less than a foot in diameter would all surely be melted away as the comet passed. No loosely packed aggregate of particles could survive. The solar wind would sweep away what was left of the comet, destroying it completely on its first passage. But Ikeya-Seki survived its last orbit relatively unscathed. And there are seven other comets like Ikeya-Seki. They form a family of sun-grazers, whose orbits follow the same pattern.

Therefore, we can assume that the nucleus of a comet has an internal strength greater than the Flying Gravel Bank model would provide.

The next try at an answer to the question was also given a vividly descriptive name: the Dirty Snowball model.

In 1950, Fred L. Whipple, of the Harvard-Smithsonian Center for Astrophysics, developed an idea first proposed by a German astronomer, Hirn, and a British astronomer, A.C. Renyard, and later elaborated on by the Soviet astronomers S.K. Vsekhsvyatsky and B.U. Levin.

Whipple described the nucleus of a comet not as a cloud of particles but as a ball of ice with dust particles trapped in it—a Dirty Snowball. A comet that was an icy mixture of ice, frozen gasses, and billions of meteoric stones and dust would start to melt as it approached the sun, releasing some of the gasses and particles trapped in the ice. It could also last through many orbits because only a thin layer of ice would be melted away on each passage.

Whipple thought the nucleus would have the consistency of "yeasty raisin bread." Such a mixture would be a poor conductor of heat; so, although the outer surface would suffer evaporation by solar radiation, the inner material would stay cold.

Whipple explained further:

> Almost all comets show peculiar random fluctuations in bright-
> ness during their flight and a very rapid brightening as they
> come near the sun. . . . The model here proposed can easily
> explain this. We can assume that a porous crust of meteoritic
> material covers much of the surface of the icy nucleus, provid-
> ing an insulating layer. As it approaches the intense heat of the
> sun, the heated nucleus tends to blow out gas and break holes
> through this crust or remove large sections of it. The underlying
> ices would produce a strong jet activity. The insulating layer
> may also account for the relatively long survival of comets after
> their capture by the inner solar system. By retarding the loss of
> their ices, the insulation may permit comets even as small as a
> mile or less in diameter to make hundreds or thousands of revo-
> lutions around the sun before they are completely dissipated.[16]

Most of the mass of a comet (about 70 percent) consists of water and
the gasses methane, ammonia, and dicyanogen, frozen into ices.
Since all of these gasses vaporize at different rates, disintegration of
the comet would be further retarded. Methane would begin to evaporate when the comet was still many astronomical units from the sun.
Next, carbon dioxide would vaporize near the orbit of Mars and,
thereafter, even closer to the sun, dicyanogen and water would become
gasses. The remaining 30 percent or so of the comet is made up of
"dust"—the heavier elements that vaporize very little even at very high
temperatures. These particles may eventually show up as meteors.

Whipple's Dirty Snowball model successfully answers many of
the questions Lyttleton's Flying Gravel Bank does not, particularly
the longevity of comets. But there are a couple of other very strong
arguments for Whipple's theory.

To begin with, many comets deviate by as much as several days
from our careful calculations of when they should reappear. It had
been assumed, before this century, that comets were delayed on their
travels by some sort of "resisting medium" in space that slowed their
motion and thus made their orbits slightly smaller. This would cause
their periods to be reduced, so that they showed up for the party early.
But when it was discovered that some comets arrive *later* than expected, the theory didn't hold up. Encke's Comet, which has the
shortest of all known periods among comets—about 3½ years—
always arrives a few hours early. And Halley's on its last apparition
was late by several days. Whereas such deviations may seem small,
astronomical observations are now so precise that errors of even this
low magnitude should be eliminated. But even when the most precise

measurements, the most powerful astronomical computers, and the most painstakingly detailed methods were used, there were still errors.

Nobody knew why their calculations wouldn't check. Until Whipple's Dirty Snowball supplied the answer.

A comet (like just about every other celestial body) rotates on its axis. Therefore the comet is constantly showing a different side to the sun. As the sun heats the exposed side of the comet, it melts the top layer and releases jets of gasses. Since the comet is constantly rotating, though, by the time the surface has been penetrated and gasses released these jets are pointing away from the sun. They now act as an engine, propelling the comet in whichever direction they are pushing, and the comet is moved ever so slightly off its course. (Unfortunately, the comet's rotation makes it impossible to calculate which way the jet stream will push the comet: It may be *either* retarded or speeded up.)

There is further proof for the Whipple theory. As we have seen (Part 1, chapter 3) on June 30, 1908, a great fireball fell out of the sky and struck the earth in Siberia. Witnesses reported that a "pillar of fire" shot up in the subarctic forest. The shock wave was strong enough to knock people down and the heat strong enough to burn the shirt off a man's back. Two villages were reported to have vanished completely, but superstitious local tribesmen were reluctant to talk about this; the devastation was immense, however. Fir trees were blown flat for an area 20 miles in diameter, and the local reindeer population was decimated.

Scientists at first presumed that a large meteor had struck the area. But when an expedition finally arrived at the site some 20 years later, no meteor or meteor fragments were found. It remained a puzzle until Whipple proposed his model. If Earth were struck by a Whipple comet the ices in the head of the comet would melt under the enormous heat of impact, and would also generate the shock wave necessary to blow the forest flat. The rest of the comet, its gasses and dust particles, would simply dissipate into the atmosphere. In no time, all trace of the comet would be gone. This is the scenario generally accepted today of what happened at the Tunguska impact site. A comet constructed along Lyttleton's Flying Gravel Bank model would have left at least some cometary fragments behind, even in the most intense explosion. The Dirty Snowball would simply have vanished.

An Estonian astronomer named Öpik came up with a slight variation on Whipple's model that he dubbed the Layer Cake model. Öpik sees a comet's nucleus as built in alternating layers of ices and dust particles. As the ices evaporate the particles are released, as in

Whipple's model. Öpik, however, also adds another layer of hydrogen at the core of the nucleus, theorizing that it was formed in the deep freeze of space and remains intact in the nucleus while temperatures remain low.

The Whipple-Öpik models are generally accepted today among astronomers and physicists as the most likely descriptions of the composition of comets.

But there are still many unanswered questions. For example, is a comet one big block of ice or a conglomerate of many smaller ice cubes? Is the nucleus spherical, and does it rotate independently? What are the surface temperatures when exposed to and when hidden from the sun? The answer to these questions will eventually solve many of the mysteries of comets—and of the universe in general.

CHAPTER 4

WHERE DO COMETS COME FROM?

After we've learned what comets are and what they're made of, the next logical question seems to be, Where do comets come from?

As we have seen, there have been dozens of theories over the centuries since Aristotle's idea that they were gaseous vapors drifting up from the earth itself, or Seneca's belief that they were just a different kind of planet. Modern scientific speculation about the origin of comets began in the 18th century.

The Capture Theory, an early idea about the origin of comets, was popular for many years. This idea states that comets travel in distant intersteller space and, by the merest of chances, wander into our Solar System and approach the sun. One of the giant planets, such as Jupiter or Neptune, then "captures" the comet by its gravitational pull and forces it to follow a new—elliptical—path. The comet will then return at regular periods of time. The Capture Theory, however, has been in disrepute for decades now. The chief problem with it was pointed out by Henry Russell of Princeton University. He showed that it was unlikely these captures had occurred within the last 10-million years. If they had, there would be many more comets with hyperbolic orbits than we now observe because there would not have been time for the planets' perturbing influence to shorten their orbits into ellipses.

The Ejection Theory started in the late 18th century. A French mathematician named Joseph Louis Lagrange hit upon the idea that comets were actually "spit out" of the giant planets such as Jupiter and Saturn. At the time, these planets were considered to be hot—sort of like miniature suns—and their internal energy was thought to be more than enough to eject a relatively small body, such as a comet, into orbit. The truth is that those planets *are* hot—but their outer gasses are very cold.

In 1870, R.A. Proctor, British astronomer, revived Lagrange's theory and worked out in detail the possible ejection of a comet from Jupiter. Proctor believed all short-period comets had their origin in Jupiter because their aphelia are usually in the area of the giant planet. His theory centered on Jupiter's Red Spot, which has fascinated scientists for centuries. Proctor considered the spot a super volcano from which the comets were blasted off the planet at a regular rate.

Proctor's theory fell by the scientific wayside until, in 1953, S.K. Vsekhsvyatsky, of the Kiev Observatory in the U.S.S.R., revived it. Vsekhsvyatsky suggested that there must be a connection because the chemistry of a comet's gasses and the chemistry of Jupiter are very similar. He believed that the short-period comets were produced by Jupiter and launched into orbit from that planet and that the longer-period comets came from the more distant planets, Saturn, Uranus, and Neptune—the longer the comet's orbit, the more distant the planet of origin.

Other scientists soon put forth objections to Vsekhsvyatsky's theory. The major problem seemed to be that the force needed to put a comet into orbit from Jupiter would be enormous. (The escape velocity of Jupiter is 37 miles per second; for Earth, it is 7 miles per second.) If the comet were ejected with enough force to escape, its momentum would carry it right out of our Solar System and we'd never see it again.

But Vsekhsvyatsky persisted. He amended his theory and suggested that the comets came instead from Jupiter's four largest moons—Io, Europa, Ganymede, and Callisto. Even though the ejection would be easier from these satellites, the astronomer failed to explain how a comet could be produced in these relatively inert worlds. The theory, therefore, remained largely ignored.

Another sort of ejection theory, widely accepted for a while, was the assertion by American astronomer T.C. Chamberlin that comets were made in the sun. The sun-grazing comets were of particular interest. It was speculated that these comets were created from the solar prominences that are thrown out at enormously high speeds from the surface of the sun itself. But the physics of the ejection itself quashed this theory. The velocity necessary to launch a comet without having it fall back into the sun would rule out any chance that it would stay within the Solar System, once ejected.

The Collection Theory was first proposed by Pierre Simon Laplace, a contemporary and countryman of Lagrange's. Laplace's idea was that comets originated in some vast interstellar cloud, a bank of nebulous matter surrounding the Solar System at about 100,000 A.U. from the sun. This cloud was captured by the sun millions of years ago, and every so often a comet coming out of the cloud at hyperbolic speed will be diverted by one of our giant planets into an elliptical orbit around the sun.

Laplace's theory was updated and expounded upon in the 1950s by Lyttleton when he was developing his Flying Gravel Bank concept. Instead of assuming that comets were already formed in the interstellar cloud, as Laplace had suggested, Lyttleton theorized that the sun, passing through the cloud on its journey around the galaxy, formed comets in the magnetic wake that trailed behind it. Many of these comets, however, formed along the so-called "accretion axis"—the direction opposite to the sun's motion. These comets would therefore escape the sun's gravitational pull; then, over time, the planets and other stars would perturb them into individual orbits. Lyttleton estimated that only a tiny percentage of the comets need to escape in this manner because "an average cloud might easily produce several thousand comets."

According to the Collection Theory, comets are being formed continuously whenever the sun moves through an interstellar cloud. That means the comet supply is being constantly renewed; so when a comet runs out of gas, so to speak, it is replaced.

To the layman, Lyttleton's theories can seem convincing. In reality, they do not hold up under mathematical scrutiny: Comets simply do not move the way they would if they had been formed in the sun's wake. Also, planets would perturb these would-be cometary particles long before they could be captured.

Oort's Cloud is today's generally accepted theory for the origin of comets. First proposed when Öpik calculated that the sun could easily maintain a family of comets as much as four light-years away, the concept was developed by Jan H. Oort, a Dutch astronomer, in 1950.

According to Oort, there are probably 100-billion or more comets in the Solar System left over from the material that first formed the planets, moons, and asteroids now circling the sun. They form a large cloud, a kind of comet storehouse, at a distance of about 100,000 A.U. from the sun—about half as far as the nearest star and well beyond the reach of telescopes. This cloud—Oort's Cloud—moves very slowly in a huge circle around the sun, taking millions of years for each orbit.

The comets remain in a fairly stable orbit until by some chance occurrence one or more of them is perturbed by a passing star and sent hurtling off into a new orbit approaching the sun. If they are further perturbed, by Jupiter or another planet, into an elliptical orbit, they are destined to become regular visitors to the sun. Such a perturbation by a passing star is thought to be the cause of the great rash of comets that visited us in the 15th and 16th centuries.

To the objection raised by astronomers that a star passing through this cloud would destroy all the comets, Öpik answered with the analogy that a star passing through such a cloud would be like a bullet passing through a swarm of gnats: The bullet would kill only a tiny fraction of the gnats without disturbing the rest of the swarm. Danger would come only if a passing star pulled the sun away from the comets. But the likelihood of such a close encounter is extremely remote. The comets that remain in the cloud have probably been there since the Solar System was formed about 4.6-billion years ago.

When comets are far from the sun they become completely inactive: The coma shrinks and the tail disappears. They go into a kind of cosmic hibernation and could then exist indefinitely in a distant cloud. Comets that are perturbed, however, begin to lose more of their gasses and meteoric particles on each successive orbit and eventually disintegrate. But the total supply of comets is so enormous, according to Oort, that the cloud has suffered very little depletion since the beginning of the Solar System.

Oort's theory is verified by many of the observed facts about comets. It explains, for example, why short-period comets are faint and long-period comets, usually, much brighter: because a comet loses a little comet material at each passage, and short-period comets come past the sun much more frequently than long-period comets. Also, it explains why bright comets are usually seen in bunches, and then not at all for long intervals: because when a star passes close to Oort's Cloud it usually perturbs many comets out of the pack at the same time.

Although Oort's Cloud is the most widely accepted theory for the origin of comets, and there is even a theory that there are two great comet clouds—one twice the distance to Neptune and the other halfway to Proxima Centauri, the nearest star—there are many skeptics. Lyttleton, for example, believes there are just not enough stars close enough to perturb anywhere near the quantity of comets that are wandering around our Solar System.

The major question Oort's theory leaves unanswered, however, is, How did the comets get there in the first place?

The comets in Oort's Cloud are either remnants of the original solar nebula or made in the cloud itself. Oort at first believed comets were originally formed near Jupiter and eventually, through such celestial mechanisms as planetary perturbation, migrated out of the cloud.

This theory, however, would require an enormous number of comets to be formed because many would be thrown out of the Solar System completely by perturbations. Also, the temperatures around Jupiter are just not cold enough to form the very complex ices of a comet. So Whipple suggested that the comets were born still farther from the sun, in the neighborhood of Uranus and Neptune. But this theory also presumes that an improbably large number of comets were originally formed.

Many scientists believe comets were formed within Oort's Cloud itself when the original gasses of the Solar System aggregated down through a complicated process. No one is sure.

But, whether comets were formed in the region of the great planets or in Oort's Cloud, most astronomers are willing to bet that comets are a by-product of the Solar System's birth. And that means that the close study of a comet can tell us much about how the whole thing began.

CHAPTER 5

THE LIGHT OF A COMET

Most people have no concept of how bright a comet is. Many believe, on the basis of stories from people who say they have seen a comet, or of what they've read, that a comet will "turn night into day." And then there are the reported sightings—of dubious merit—from the Renaissance and before. It's hard to trust some of these early observers, no matter how zealous they were as sky watchers, who (for example) reported in 1066 that Halley's Comet "lighted up the entire night."

Actually, very few comets are as bright as the brightest planets. In fact, the comets that could become very bright usually pass so close to the sun that their display is hidden.

Magnitude is a scale developed by astronomers to measure the brightness of celestial objects. The *magnitude* of a comet, star, or planet is a measurement of how bright it is. The lower the number, the brighter the object: A star of magnitude 4 is brighter than one of 9. An object of magnitude −1 is brighter still. A difference in magnitude of 5 is exactly 100 times brighter. That is, an object of magnitude −10 is 100 times brighter than an object of magnitude −5. Sirius, the brightest star, is of magnitude −1.4. Venus is often as bright as −4 or more; the full moon is magnitude −12. The sun is −24.

Because comets do not give one bright point of light (the way a star or planet does) but more a blur of light spread over a wide area, it is difficult to estimate magnitude. Most such estimates are measures of the brightest part of the comet—the nucleus. The Great Comet of 1744 was measured at −1 magnitude. The Comet of 1882 I and the

Comet of 1965 VIII were between −10 and −15 in magnitude; in other words, they were very bright objects. Comet West of 1976 was of a −3 magnitude; Ikeya-Seki of 1963 was about magnitude 3. Most normal long-period comets are in the 7–10 range of magnitude; however, most comets fall between magnitudes 4 and 6.

Although a comet may at times appear starlike, it is not luminous. Instead, like every other object in our Solar System, it shines because of the sun. Most of a comet's light comes from the sun's rays bouncing off the scattered dust particles in the coma and tail. These small but solid particles reflect and scatter the sun's light much the way the particles in the rings of Saturn do.

A comet, however, unlike planets, asteroids, or moons, does not glow by reflected light alone. Much of its brightness depends on fluorescence, a process in which the gas molecules of the comet are excited by sunlight and thereby create their own light: The sunlight absorbed by the molecules at one wavelength is emitted at another, giving off a strong light similar to that produced by the excitation of gasses in a neon light tube.

Because of its complicated light sources, a comet's brightness is often very difficult to predict. Many comets that, when first spotted, promise to give a spectacular show turn out to be monumental disappointments. Kohoutek's Comet of 1973 was the most recent—and glaring—example of this. Ballyhooed as easily the brightest comet yet seen this century, when it arrived it wasn't even visible to the naked eye!

The problem of predicting a comet's brightness is complicated by the sun-grazing comets, which should be very bright. These comets come so close to the sun that in fact their brightness is lessened because the area of the sun's surface the comet is exposed to is actually diminished. Contrary to the standard formula—that a comet gets brighter the closer it gets to the sun and fades as it draws away—a comet can, therefore, actually be reduced in magnitude as it nears the sun. Ikeya-Seki I of 1968 and Giacobini-Zinner of 1972, for example, are comets whose magnitude actually diminished when they reached perihelion.

There are other comets whose brightness is unpredictable because their magnitude increases without warning. Holmes's Comet of 1892 increased by a magnitude of nine in one huge, unexplained leap as its coma suddenly grew larger than the sun. This behavior was so sudden, and so uncharacteristic, that several astronomers thought the comet was not a comet at all but instead a dust cloud left behind by the collision of two planets. However, Holmes's Comet has made several returns since 1892; so that theory has been laid to rest.

Like many comets known as "sun-grazers," the Comet Ikeya-Seki of 1965 produced a long, spectacular tail. *Lick Observatory photograph.*

The Comet Schwassmann-Wachmann I often makes sudden jumps in magnitude of as much as five to nine points. First, its nucleus suddenly gets very bright, and soon after the comet grows very large, and then follows the sudden jump in magnitude. Some astronomers have tried to link such behavior by this and other comets to outbursts of sunspot activity; but no positive correlation has been established. Others have tried to suggest that perhaps these oddballs aren't comets at all but rather tiny planets composed of antimatter. Then, when meteors hit the surface of these comet-planets, the result is a nuclear explosion that accounts for the sudden change in brightness.

CHAPTER 6

LOST COMETS, DEAD COMETS, AND METEORS

If the comet's tail is the visible trail of debris from its head, it stands to reason the comet will eventually run out of stuff to feed the tail. There is no doubt that a comet loses a great deal of its material every time it passes perihelion. The only question seems to be, How long will it take?

On a cosmic scale, in fact, the life span of a comet is shorter than a fruit fly's. Most celestial bodies have enormous life spans. Earth has been around for 4.6-billion years. The sun is probably older than that. And, from the look of things, it's a pretty good bet that the Solar System will last a good deal longer. But comets are short-lived. We know that at least several die each century.

In the case of very-long-period comets, their visits to the sun are so rare that their birth most certainly preceded Mankind's appearance—and they may outlast us altogether. But those comets that make frequent passages of the sun aren't long for this world.

About 1 percent of a comet's mass is lost during each orbit. A comet with a diameter of 1 mile loses a 25-foot-thick layer of dust and ice on each fly-by. After a time, after all the ices are stripped away, all that will be left is a conglomerate of dust and rock, which will soon break up. The comet is dead.

This is why short-period comets, which in all likelihood have made many more trips around the sun, are usually not as bright as long-period comets and certainly die out sooner. The comet with the shortest known period, Encke's Comet, comes to perihelion 30 times

The Comet West of 1976 showed a tail of multiple streamers, which often signals the breakup of a comet's nucleus. *Lick Observatory photograph.*

each century. It was identified almost two hundred years ago, and some astronomers now estimate it will make its last passage sometime between the years 1990 and 2000.

There have been many instances in this century of comets fading away. On November 13, 1974, John Bennett, a noted amateur comet hunter, found a bright comet that seemed to be growing in magnitude by the day. The perihelion was calculated for two weeks later. The comet was expected to be quite visible to the naked eye by November 22, but as it approached the sun its intensity weakened and the head became diffuse. On November 25, when it should have been at its brightest, it disappeared altogether. Two weeks later, on December 8 and 10, the astronomers C. Torres and K. Parra in Chile recorded the comet's "ghost" on photographic plates: All that remained was "a faint nebulous mass." The comet had simply melted away.

There are many other examples of comets' expiring. Alcock's Comet of 1959 passed the sun a fairly bright comet—and within a week faded away. The Comet Daido-Fujikawa vanished after it rounded the sun in 1970. Comet Westphal and Comet Finsler suffered similar fates, also in this century.

Will Halley's Comet die, perhaps, on this 1986 passage of the sun?

That is highly unlikely. Like all comets, Halley's is slowly disintegrating; but its resources are large. Most important, Halley's Comet just doesn't make that many orbits of the sun in each century. A visit once every 76 years or so does very little to drain its total reserves.

Some astronomers have reported that Halley's has diminished in brilliance during its last few passages. Yet these reports seem exaggerated. From A.D. 837 to 1910, its brightness remained at about the same level. The length of its tail display also remained very much the same. Just how brilliant Halley's will be in 1986, no one can say for sure. But any comet that's reappeared regularly for 2000 years has got to be considered reliable.

One of the more dramatic ways for a comet to die is to have its nucleus split into two or more parts.

One of the most striking events of this kind happened to Brooks's Comet. Seen first in July 1889, it had made a very close approach of Jupiter three years earlier, actually passing between its moons at less than 0.25-million miles from the giant planet. The enormous gravitational pull of Jupiter ripped Brooks's into five parts, a main nucleus followed by four smaller pieces. Two of these pieces soon disintegrated. Then a third piece expanded and faded away. The fourth piece lasted longer, growing its own tail, but then it, too, disintegrated.

Brooks's Comet returned on schedule in 1896, but now it was a much reduced spectacle. This faint comet has returned as predicted ever since, but it is not expected to last out this century.

Probably the most famous case of a comet's breaking up was Biela's Comet. In 1772, Charles Messier discovered a relatively faint comet. In 1819 the comet was spotted again, by Friedrich Bessel, who confirmed it was the comet Messier had seen in 1772. Its period was calculated at 6 years and 9 months, and Bessel predicted its return in 1826.

On February 27, 1826, an Austrian army captain named Wilhelm von Biela spotted the comet, and the observation was confirmed by the French astronomer Adolphe Gambart. It showed up again in 1832, but was missed, because of a poor position in the sky, in 1839.

On November 28, 1845, Biela's was seen again—this time by

the astronomer De Vico in Rome, who reported nothing strange. On December 19, though, the comet was suddenly reported to have developed a pear shape, and ten days later it split in two. When Professor James Challis at Cambridge first spotted the two comets, he didn't believe his eyes and dismissed the sighting as an instrument error:

> On the evening of January 15 [1846], when I went down to observe it, I said to my assistant, "I see *two* comets." However, on altering the focus of the eyeglass and letting in a little illumination, the smaller of the two comets appeared to resolve itself into a minute star, with some haze about it. I observed the comet that evening but a short time, being in a hurry to proceed to [other] observations. [17]

But Matthew Fontaine Maury couldn't dismiss what he'd seen. He observed the comet 49 different times on this passage and was given credit for discovering the split. John Herschel described what the astronomers saw:

> It was actually seen to separate itself into two distinct comets, which, after thus parting company, continued to journey along amicably through an arc of upwards of 70 degrees of their apparent orbit, keeping all the while within the same field of view of the telescope pointed towards them. [18]

The smaller of the two comets followed the other at a distance of about 165,000 miles. Each comet developed its own faint tail.

In February 1846 Maury reported that he saw a bridge of light arc from the larger comet to the smaller one. Eventually this formed into one long tail, and the larger comet developed two other tails.

By now, Challis had changed his mind about what he'd seen:

> There are certainly two comets. The north preceding is less bright and of less apparent diameter than the other, and has a minute stellar nucleus . . . I think it can scarcely be doubted, from the above observations, that the two comets are, not only apparently but really very near each other, and that they are physically connected. When I first saw the smaller, on 15 January, it was faint, and might easily have been overlooked. Now it is a very conspicuous object, and a telescope of moderate power will readily exhibit the most singular celestial phenomenon that has occurred for many years—a double comet. [19]

Father Angelo Secchi of the Vatican observatory was the first to spot Biela's double comet on its next return, in 1852. The two comets were now separated by 1.5-million miles, and no one could tell which was the original and which the offshoot.

In 1859 the position of the comets was bad for viewing and, as expected, they were not observed. In 1866, however, when they should have been easy to spot, again they were not observed. And they have never been seen since.

Biela's Comet had surely died. But did it leave a corpse?

Many people confuse "shooting stars"—meteors—with comets. A comet lies well beyond the earth's atmosphere, and because of its great distance, takes many weeks or months to move across our sky. A meteor, on the other hand, shoots across in a fraction of a second and burns up within the earth's atmosphere.

A meteor is a particle—usually about the size of a grain of sand—that orbits the sun. If it drifts to within 120 miles of Earth it enters our atmosphere. As it bangs its way through the gas molecules of the upper atmosphere at speeds exceeding 25 miles per second, friction heats it until it is red hot. The meteor literally burns itself up, and we on Earth may glimpse it as that bright streak of light we call a shooting star.

Few meteors ever reach the surface of Earth. Most vaporize in the upper atmosphere, although some meteor dust does settle down to the earth. In fact, 150-million meteor particles fall on the earth each year—amounting to somewhat less than a ton of dust.

Meteors also come in great "showers" or "streams." There are many different showers, and each comes from a particular area of the sky and at a particular time of year. The August shower comes from the vicinity of the constellation Perseus and is known as the Perseid stream. The October shower is in the area of Orion; so these meteors have been dubbed the Orionids.

Here are the major meteor showers and the times of year when they occur.

Quadrantids	January
Lyrids	April
Eta Aquarids	May
Delta Aquarids	July
Perseids	August
Orionids	October
South Taurids	November
Leonids	November
Geminids	December
Ursida	December

These showers are more spectacular in some years than in others. This indicates that the cosmic dust clouds that cause the displays are in orbit around the sun: The meteor material spreads out along this orbit, and when the earth happens to cross the orbit of the meteor stream, the display—the shower we see—becomes particularly impressive.

It was not until the mid-19th century that scientists discovered the link between comets and meteors. In 1866 an Italian astronomer, Giovanni Schiaparelli (who is best known for discovering the "canals" of Mars), proved that the Perseid meteor shower follows the same orbit as Comet Tuttle. He computed the dimensions and position in space of this group of meteors and found they matched exactly those of a bright comet which had passed the earth in the summer of 1862.

Schiaparelli's findings were soon confirmed when the November meteor showers were shown to follow the same orbit as a comet of 1866.

Since the fragments of most meteor showers are, like a "dead" comet, composed of a loose swarm of small particles scattered over a large area along an orbit path, it took a very small leap of the imagination to tie the two together. This assumption has since been proven beyond doubt.

That brings us back to the mysterious disappearance of Biela's Comet. Once Schiaparelli identified Comet Tuttle with the Perseid meteor stream, other astronomers began trying to link other meteor streams with comets.

Two German astronomers, Ernst Weiss and H.L. d'Arrest, rightly supposed that the death of Biela's Comet may, in fact, have given life to a meteor stream. In 1867, after the double comet missed its expected 1866 appearance, they suggested that the Andromedid meteor shower was probably connected with that comet. Another shower was predicted for 1872, to coincide with the schedule Biela's was on before it split up. The shower showed up right on time. It was reported that more than 75,000 shooting stars per hour were seen.

Over the next three decades, the shower, now called Bieliid, became gradually weaker until it petered out in 1899. A few meteors are still seen each year; but, practically speaking, Biela's Comet has finally died.

Other meteor showers are now known to be associated with comets. The Beta Taurids, which occur between late June and July every year, are linked with Encke's Comet. The Leonids are connected with the Comet of 1866 I and occur every 33 years, but they are not reliable, usually because of planetary perturbations.

Halley's Comet is associated with two meteor displays—the Eta

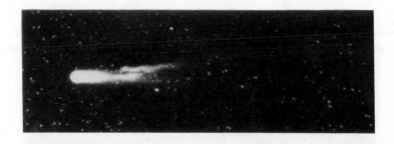

Halley's Comet is associated with two meteor showers. Note the dust particles separating from the tail of the comet. They may someday show up as meteors. *Lick Observatory photograph.*

Aquarids, which show up between late April and mid-May, and the Orionids of late October.

Comets don't just spawn meteors. It is well known that comets also contribute millions of tiny particles to the Solar System as they move through interplanetary space. Some of this other comet debris is seen as the zodiacal light—a visible glow in the eastern sky before sunrise and the western sky after sunset, especially near the equator. Comet dust also contributes to the so-called ''counterglow''—a faint glow in the night sky in the direction opposite to the sun.

In order to maintain these atmospheric effects, comets must spew into the Solar System some ten tons of dust per second. Otherwise, solar wind and natural disintegration would sweep the atmosphere clean of all comet dust.

PART FOUR

THE RETURN OF HALLEY'S COMET

CHAPTER 1

HALLEY'S COMET THROUGHOUT HISTORY

Fortunately for future astronomers, Chinese and Japanese scholars viewed comets as emissaries from the different peoples of the earth and as forecasters of important events. Therefore, these ancient Oriental wisemen kept careful track of the comings and goings of comets: the dates of their appearances and the paths each took.

Because of these meticulous records, present-day astronomers have been able to trace the path of Halley's and other comets right back to antiquity. We know of at least 29 times Halley's Comet has orbited the sun. The earliest definite date is 239 B.C. It almost certainly made another appearance in 163 B.C., but nobody seems to have spotted it. But the next two occasions were recorded by the Chinese in 83 B.C. and 11 B.C.

All subsequent appearances have been definitely verified. In A.D. 141, the Chinese saw a comet in the spring that was "six or seven cubits long" and of a bluish white hue. In A.D. 218, Dion Cassius, a Roman, described Halley's Comet of that year as "a fearful star whose tail flew from west to east." In 451, a comet seen over Europe was given credit for the victory of the Roman general, Aetius, over Attila the Hun at Châlons. In 760, a comet "like a great beam" was observed during Emperor Constantine's reign.

It was in 1066 that Halley's Comet made its most famous return, when (as we have already seen) it accompanied the great Norman Conquest of England and the Battle of Hastings. A contemporary Greek historian, Zonares, wrote that Halley's Comet on that return

111

was as large as the full moon. The 1222 appearance of Halley's was said to presage the death of King Philip Augustus of France; meanwhile, in China a scholar named Ma-tuaon-lin stated that it was 30 cubits long and died out in two months.

The Great Comet of 1456 was said by the Chinese to have a tail like a peacock's and a head the size of a bull's eye. Then came the Comet of 1531, observed by Apian, and that of 1607, seen by Longomontanus—both mentioned in Halley's *Synopsis*. At its next appearance, in 1682, Halley's Comet was greeted by Edmund Halley— and the rest is history.

In 1835 Halley's Comet passed around the sun five days later than its predicted perihelion. But by 1910, the advancement in scientific techniques and instruments made a more exact prediction possible. A cash prize was offered to the astronomer who could fix most accurately the comet's perihelion. Two British astronomers at Greenwich, P.H. Cowell and A.C.D. Crommelin, came within three days of an accurate prediction and won the award. They had traced the comet's path from its perihelion in November of 1835 backward to 1759 and forward to its 1910 passage. Their prediction proved to be so accurate that when the comet was first spotted it was said to be only yards from where Crommelin and Cowell had said it would be.

It was Max Wolf of the Heidelberg Observatory in Germany who first detected Halley's Comet on September 12, 1909, on a photographic plate. (It was later realized that the first photograph of Halley's was actually taken at Helwan Observatory in Egypt on August 24, but Wolf was the first one to identify the celestial object as the comet.) The comet was still more than 300-million miles away from perihelion, out beyond the orbit of Mars. It looked no different from all the faint stars it was among and was only identified because of its night-by-night movement.

When Halley's Comet came within 14.3-million miles of Earth on May 20, 1910, it was traveling at a relative speed of 52 miles per second. (Because Halley's Comet and the planet Earth are orbiting in opposite directions, and Earth's speed is 19 miles per second, Halley's actual speed in 1910 was nearer to 32 or 33 miles per second.)

On May 18 the comet passed directly between Earth and the sun. No trace of the comet could be seen while it was in front of the sun, confirming the fact, already well established, that a comet is "a big bag of nothing."

On May 21 Earth passed through the tail of the comet with no ill effects, although many people had predicted that the inhabitants of Earth would be wiped out by cyanogen gas detected in the nucleus of the comet.

Halley's Comet as it appeared in 1910.

Astronomers were able to follow Halley's photographically until June 1911, when it was more than 500-million miles away from the sun and out beyond the orbit of Jupiter.

Perhaps the finest description of the 1910 appearance of Halley's Comet was given by Mary Proctor in her book, *The Story of Comets.*

113

Here is the author's account of her view from the roof of the Times Building in New York City:

> Then came May 4, a bitterly cold morning; but the stars shone brightly and there was every hope of the comet being visible from the tower heights. These hopes were confirmed, for on stepping out on to the parapet the writer saw the comet in all its splendor . . . with a head shining as a star of about the second magnitude, and surrounded by a nucleus. Extending outward like the beam of a searchlight gleamed the tail nearly fifteen degrees in length. Calling down to the janitor to make known the good news, the balcony was soon filled with eager members of the Times staff. . . . Spurts of light like tiny waves seemed to flow out from the nucleus to a distance of two or three degrees. At twenty minutes to four, the writer, on looking downward at the horizon, was startled by what appeared to be a streak of flame, but as it rose higher it proved to be the crescent moon, which with the comet and the planet Venus, completed a wonderful trio. The comet remained visible, resembling a bright star with a slender stream of silvery mist trailing a few degrees after. By four o'clock it had faded in the light of approaching dawn.[20]

For the 1986 return of Halley's Comet, a worldwide cooperation called "The International Halley Watch" has been organized by NASA, the National Aeronautics and Space Agency of the United States. All information and observational data from astronomers throughout the world will be coordinated by the Halley Watch through the Jet Propulsion Laboratory in Pasadena, California, and detailed instruction for viewing of the comet will be issued by J.P.L.

Astronomers at Palomar Observatory in Pasadena, northeast of San Diego, California, first detected Halley's on October 16, 1982, heading toward the sun. It had passed aphelion 3.3-billion miles from the sun, well beyond the orbit of Neptune, back in 1948. In 1964 it passed near Neptune's orbit on its way back toward the sun, and it was halfway between Neptune and Uranus by 1974. It passed the orbit of Saturn in 1984 and will first pass Earth—but not close—in November 1985.

Amateurs' telescopes should be able to pick out Halley's Comet as soon as September or October of 1985, and it will be visible to the naked eye no later than March 1986.

Perihelion is predicted for 3:50 A.M. (Eastern Standard Time) on February 9, 1986, when the comet will be only 56-million miles from the sun. It will pass closest to the earth on April 11, 1986—about

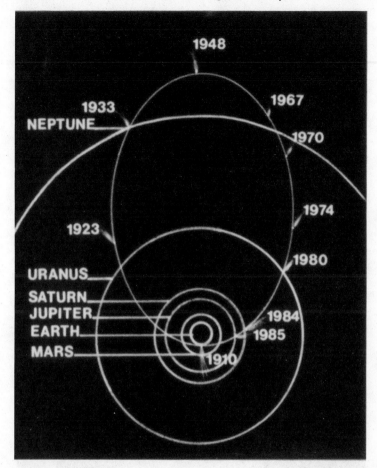

The orbit of Halley's Comet from May 1910 to its perihelion in February 1986.

39-million miles away from us—on its way back after perihelion.

As the earth and the comet change positions in their ever-changing orbits, the observing conditions will change. Using the calculation of orbital and magnitude estimations made by astronomers Donald Yeomans and Brian Marsden of the Jet Propulsion Laboratory, an amateur with a reliable telescope can expect to pick up Halley's Comet around August of 1985. It will then be at about magnitude 14 and visible in the eastern sky before dawn.

In late October, at about magnitude 10, the comet will be visible near the Crab Nebula and in November near the Pleiades. In mid-November, when it is just passing Earth on its way toward perihelion it will have brightened to magnitude 8.

By December, Halley's will have begun to develop its first wisp of a tail and later that month, at magnitude 6, it will be visible with a good pair of binoculars.

The first naked-eye sightings of Halley's should occur sometime in early January. It will be visible in the low western sky, along with Jupiter, later in the month; at magnitude 4 or 5, the comet, its tail now growing to substantial length, should be quite impressive when viewed through binoculars.

At perihelion—during the second week of February—the comet will all but disappear in the twilight as it approaches and passes behind the sun.

Once it emerges from behind the sun, Halley's will be visible in the morning sky an hour before dawn. It will be about magnitude 2 and the tail will have grown in length by three or four times—it should be impressive to the naked eye. Each day the comet will appear earlier, as it moves south in the sky. By mid-March the tail will stretch out over at least one sixth of the sky.

The length of the tail and the brightness of the comet should re-

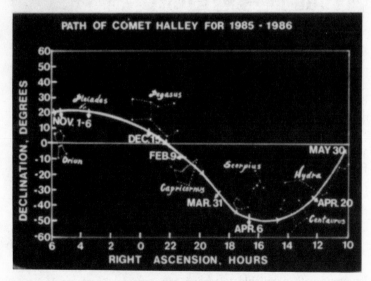

The path Halley's Comet will take through the night sky from November 1985 through May 1986.

main steady through April, when it again passes close to the earth. By the end of the month the comet will be near the constellation Leo and may be in the best position for viewing from the northern hemisphere. (Australia and Argentina in the southern hemisphere will probably get the best view, however, as Halley's will be directly overhead in the dark night sky.)

In early May 1986, Earth will experience a meteor shower of debris from Halley's, although the comet will have passed the same area seven weeks before. Thereafter, Halley's will move northward in our sky and begin to fade quickly from about magnitude 7 in May to 11 in July and, soon after, will be lost to the sight of most amateurs, although professional astronomers will continue to track the comet through 1987.

Just how bright will Halley's Comet be in 1986? Well, most experts predict that it will not be at its most brilliant. Its unfavorable position in the sky and its relatively far-off approach to the sun and the earth will diminish its display. Then, too, Halley's Comet may be getting old and have suffered enough wasting of its nuclear material to affect its magnitude.

In addition, the planet Earth is getting more and more polluted by both light and waste, and so if you are trying to view the comet from a heavily populated area you may be disappointed. The best visibility will come where there are clear and dark skies. A little vacation out to the country to view Halley's Comet will be worth the trip.

If you plan to use a telescope or high-powered binoculars to view Halley's Comet it would be best to first read one of the many well-written books on amateur astronomy. *The Amateur Astronomer's Handbook* by J.B. Sidgwick (Dover Publications, New York, 1980, $6.95) and *Skyguide* by Mark R. Chartrand (Golden Press, New York, 1982, $6.95) are two of the best. Both books contain complete instructions on how to use telescopes and binoculars and how and when to best view comets and all the other celestial objects in our sky. Also, both give helpful tips on purchasing astronomical equipment for those who have never owned their own telescope or binoculars suitable for viewing the heavens.

In addition, the August 1984 issue of *Science Digest* magazine contains a special section for "Amateur Astronomers," including "How to Buy a Telescope," "How To Become a Stargazer," and there's even an article on how to take celestial pictures.

One final note. Unlike solar eclipses, there is absolutely no danger in viewing the comet with the naked eye. So whether you have sophisticated telescopic equipment or, like human beings have done throughout the ages, you plan on simply viewing with the unaided eye, Halley's Comet will, with any luck, offer you a once-in-a-lifetime spectacle.

Whether you view the comet with binoculars or a small telescope or your naked eyes, it's best to find a spot away from city lights and pollution. Also, at night it's important to allow the eyes at least 10 to 20 minutes to adapt to the darkness. Going from a brightly lit indoor environment to the darkness of outdoors to look for the comet will make it difficult to locate, and impossible to see well.

A skymap is exactly like a map of the earth laid out flat. The 0° declination line (see ''Path of Comet Halley for 1985–1986'') marks the separation between the northern and southern hemisphere skies. When the comet's path drops below the 0° mark it will be primarily in the southern hemisphere. Since a comet is of such large dimensions, however, portions of the tail will, of course, still be visible to North American viewers.

To use the skymap, locate a well-known constellation such as Pegasus (Ursa Major, the ''Big Dipper,'' probably the most familiar constellation to the unexperienced sky watcher, is not pictured in this skymap. Its location would be just off the page in the upper right hand corner of the map). Now line the skymap up according to the placement of the stars. The skymap on page 116 traces the path of the comet in the sky from November 1985 to May 1986. During this period the comet goes from being primarily a northern hemisphere object in late 1985 to a southern hemisphere one in early 1986.

CHAPTER 2

SPACE MISSIONS TO HALLEY'S COMET

With Halley's Comet about to make its appearance, scientists are preparing the biggest coordinated space effort ever undertaken to study a celestial object. Several nations plan to send space probes to give Mankind its best look yet at a comet. Plans now call for a very sophisticated spacecraft rendezvous with the comet by the European Space Agency (E.S.A.) under the project name *Giotto,* a Soviet fly-by, and a Japanese mission.

The ideal comet for such close-up inspection is a long-period one that has visited the Solar System infrequently since it left the Oort Cloud. But, because the appearance of such a comet is unpredictable, plans for a space mission and launch would be impossible. The brightest predictable comet that hasn't worn itself thin, one that still has an ample supply of gasses and dust typical of a new comet, is Halley's. All other comets of this type either are too dim or will not return this century.

The use of three spacecraft in this international endeavor shows a degree of cooperation never seen before in space exploration. Heads of all three projects have met regularly for several years now. But this cooperation and communication is necessary if the missions are to succeed in gathering the mountains of information due to be revealed.

Each spacecraft will have scientific instruments of all types: cameras and infrared cameras, plasma and dust detectors, neutral gas detectors and magnetometers. Measurements on one probe will assist and complement those from another.

119

The *Giotto* space probe will penetrate into the coma of Halley's Comet, sending back information to Earth before it is destroyed by comet dust. *Lick Observatory photograph.*

The mission planned by the E.S.A. is the most advanced. It is named for Giotto di Bondone, the Florentine painter who in 1301 painted a fresco of Halley's Comet on the walls of the Scrovegni chapel in Padua. In July of 1985 the 0.75-ton probe will be launched aboard an Ariane spacecraft from a complex in French Guiana. The probe will stay in an elliptical orbit for eight months. Then, on March 13, 1986, it will rendezvous with Halley's Comet at about one A.U. from the earth.

Flying at tremendous speed, the probe will dive like a kamikaze through the comet head. Blasted by comet dust, it will be lucky to last for an hour or so as it flies within half a mile of the nucleus gathering the information the scientists are after. It probably will not survive the trip. In order to save the gathered data, *Giotto* will transmit information as it receives it, rather than storing it on tape (as most other probes do). In order to save it from instant destruction by the comet itself, however, *Giotto* has been equipped with a cosmic "chest protector"—dust shields designed to block oncoming comet particles. The shields are made of 1-millimeter-thick sheets of aluminum backed by Kevlar, the material used in bullet-proof vests. Yet in the end, the enormous amount of dust particles in the coma will almost certainly destroy the spacecraft.

The goals of the *Giotto* mission include: determining what the nucleus and coma are made of, getting good pictures of the nucleus, and discovering the composition of the ionized gasses in the coma and how they react to the solar wind.

A second mission to Halley's Comet will be launched by the Institute of Space and Astronautical Science (ISAS) in Japan. It is designated *Planet A* and will be launched from the Kagoshima Space Center in southern Japan on August 14, 1985. It will intercept Halley's five days before *Giotto* does, on March 8, 1986.

Only two instruments are aboard *Planet A*—an ultraviolet camera and a solar wind analyzer—and the probe is only designed to observe the growth of Halley's coma, the speed of the hydrogen atoms around the nucleus as they expand out through the coma and the tail, and the magnitude of the solar wind.

This probe will not venture as close to the comet as *Giotto,* and so it will not be taking close-up shots of the nucleus; but it will photograph the coma around it.

Vega, to be sent up by the U.S.S.R., will actually be two spacecraft—one to be launched on December 22 and the second on December 26, both in 1984. The spacecraft will first fly by Venus, sending probes to that planet in June of 1985. Vega 1 and 2 will then head for an encounter with Halley's Comet.

Vega 1 will rendezvous with the comet on March 8, 1986, and Vega 2 will arrive a week later. Either probe can be sent to within 50 miles of the nucleus of the comet, but experts believe the Soviets will only permit the second probe to attempt this close approach because, like *Giotto,* it will almost certainly be destroyed. Vega 1 will be saved to collect data and will probably get no closer than 5000 miles from the nucleus.

The Soviet Union's probes will carry spectrographic equipment that will provide spectral data of the nucleus as well as infrared pictures. These observations will provide information helpful in figuring the temperature, nature, and content of the comet's nucleus, dust particles, and gasses.

The United States has decided that a probe of Halley's Comet would be too expensive. Instead, it has been decided to send the International Sun-Earth Explorer satellite, already in orbit since 1978, to rendezvous with another comet first. The gravitational pull of the moon will catapult the satellite to within 3000 miles of Comet Giacobini-Zinner in 1985 and then on to a long-distance look at Halley's Comet in 1986.

The Explorer will fly through the tail of Giacobini-Zinner on September 11, 1985. Then it will approach to within 20-million miles of Halley's, recording data about the comet and the solar wind.

CHAPTER 3

HALLEY'S COMET EVENTS, 1985–86

The following planetariums and observatories in the United States are planning many sorts of programs to accompany the return of Halley's Comet in 1985–86, including lectures, space shows, public viewings, and information centers. Call the planetarium in your area for a complete schedule of Halley Events, starting in the fall of 1985.

If none of the observatories listed below are in your area, you can get a more complete list from the book, *U.S. Observatories: A Directory and Travel Guide,* by H.T. Kirby-Smith (published by Van Nostrand Reinhold Co., New York, 1976).

[One other note: Although plans for network television and PBS to broadcast special programs and intensive coverage of Halley's have not yet been announced, there is sure to be at least some programming devoted to this historic occasion. So be sure to check your TV listings and watch for announcements of coming programs on Halley's Comet 1986.]

EAST

Allegheny Observatory
Department of Physics and Astronomy
University of Pittsburgh
Pittsburgh, PA 15214
(412) 624-4290

American Museum-Hayden Planetarium
81st Street at Central Park West
New York, NY 10024
(212) 873-8828

Buhl Planetarium
Allegheny Square
Pittsburgh, PA 15212
(412) 321-4300

Davis Planetarium
Maryland Science Center
601 Light Street
Baltimore, MD 21230
(301) 685-2370

Albert Einstein Spacearium
National Air and Space Museum
6th and Independence Avenue
Washington, DC 20560
(202) 381-4193

Fels Planetarium
20th and Parkway
Philadelphia, PA 19103
(215) 448-1292

Harvard-Smithsonian Center for Astrophysics
"Open Nights" Center for Astrophysics
60 Garden Street
Cambridge, MA 02138
(617) 495-7000

Charles Hayden Planetarium Museum of Science
Boston, MA 02114
(617) 723-2500

Sproul Observatory
Swarthmore College
Swarthmore, PA 19081
(215) 544-7900 Ext. 207

Strasenburgh Planetarium
663 East Avenue
Rochester, NY 14603
(716) 244-6060

United States Naval Observatory
Washington, DC 20390
(202) 254-4533

Vanderbilt Planetarium
178 Little Neck Road
Centerport, NY 11721
(516) 757-7500

SOUTH

R. C. Davis Planetarium
P. O. Box 288
Jackson, MS 39205
(601) 969-6888

Fernbank Science Center Planetarium
156 Heaton Park Drive
Atlanta, GA 30307
(404) 378-4311

Louisiana Arts and Science Center Planetarium
502 North Boulevard
Baton Rouge, LA 70801
(505) 344-9465

Morehead Planetarium
University of North Carolina
Chapel Hill, NC 27514
(919) 933-1237

National Radio Astronomy Observatory
Charlottesville, VA 22901
(804) 296-0211

Space Transit Planetarium
3280 South Miami Avenue
Miami, FL 33129
(305) 854-4242

MIDWEST

Talbert and Leota Abrams Planetarium
Science Road
Michigan State University
East Lansing, MI 48824
(517) 355-4673

Adler Planetarium
1300 South Lake Shore Drive
Chicago, IL 60605
(312) 322-0304

McDonald Observatory
Department of Astronomy
University of Texas
Austin, TX 78712
(512) 471-4462

McDonnell Planetarium
5100 Clayton Avenue
St. Louis, MO 63110
(314) 933-1237

Warner and Swasey Observatory
1975 North Taylor Road
East Cleveland, OH 44112
(216) 451-5624

Yerkes Observatory
University of Chicago
Chicago, IL 60637
(312) 753-8180

WEST

Grace H. Flandrau Planetarium
University of Arizona
Tucson, AZ 85721
(602) 626-4515

Reuben H. Fleet Space Theater
1875 El Prado
San Diego, CA 92103
(714) 238-1233

Gates Planetarium
Colorado Boulevard and Montview
Denver, CO 80205
(303) 388-4201

Griffith Observatory
2800 East Observatory Road
Los Angeles, CA 90027
(213) 661-1181

George T. Hansen Planetarium
15 South State Street
Salt Lake City, UT 84111
(801) 364-3611

Hale Observatories (Mt. Wilson and Mt. Palomar)
California Institute of Technology
Pasadena, CA 91101

Kitt Peak National Observatory
P. O. Box 26732
Tucson, AZ 85726
(602) 325-9204

Lick Observatory
Department of Astronomy
University of California at Santa Cruz
Santa Cruz, CA 95064
(408) 429-2513

Morrison Planetarium
Academy of Sciences
San Francisco, CA 94118
(415) 221-5100

FOOTNOTES

1. Henry Pemberton, *A View of Sir Isaac Newton's Philosophy* (London: 1728), unpaginated preface.
2. W.H. Turnbull and J.F. Scott, eds., *Correspondence of Isaac Newton* (London: Cambridge University Press, 1960), 431.
3. Frank E. Manuel, *A Portrait of Newton* (Cambridge: Belknap Press, 1968), 159.
4. J.L.E. Dreyer, *A History of Astronomy From Thales to Kepler* (New York: Dover Press, 1953), 353.
5. H.A. Howe, *Elements of Descriptive Astronomy* (New York: Silver, Burdett and Co., 1897), 189.
6. E.F. MacPike, *Correspondence and Papers of Edmond Halley* (London: Taylor and Francis, 1932), 49.
7. Ibid., 50–51.
8. J.R. Hind, *The Comets* (London: J.W. Parker and Son, 1852), 50.
9. Ibid., 41.
10. Mary Proctor, *The Romance of Comets* (New York: Harper Brothers, 1926), 99.
11. Augustus De Morgan, *Essays on the Life and Work of Newton*, P. Jourdain, ed. (London: Open Court Publishing Co., 1914), 384.
12. MacPike, 109.
13. *Biographia Britannica* (London: Britannica, 1757), iv., 2516.
14. John C. Brandt and Robert D. Chapman, *Introduction to Comets* (New York: Cambridge University Press, 1981), 12.
15. John C. Brandt, *Comets* (San Francisco: W.H. Freeman and Co., 1981), 82.
16. Fred L. Whipple, "Comets," *Scientific American* (July, 1951): 22.
17. Proctor, 53.

18. J.F.W. Herschel, *Outlines of Astronomy* (London: Longmans and Green, 1871), 390.

19. Patrick Moore, *Guide to Comets* (London: Lutterworth Press, 1977), 86.

20. Proctor, 109–110.

GLOSSARY

Aphelion: The point in a comet's or a planet's orbit when it is farthest from the sun.

Astronomical Unit (A.U.): One A.U. equals the distance between the earth and the sun—about 93-million miles.

Coma: The bright, cloudlike material surrounding the nucleus of a comet, forming the head.

Ellipse: The football-shaped closed curve of the orbit of a body orbiting the sun.

Head: The part of a comet that includes the nucleus and the coma.

Hyperbola: An open curve. A comet following this orbit would never return to the sun.

Meteor, or shooting star: A small orbiting particle, usually no bigger than a grain of sand, that shows a bright streak of light when it enters the earth's atmosphere and burns itself up.

Nucleus: The densely packed center of a comet's head.

Parabola: An ellipse stretched out to infinity. A theoretical concept, rarely seen in nature.

Perihelion: The point in a comet's or a planet's orbit when it is closest to the sun.

Period: The time it takes an orbiting body to make one complete orbit.

Perturbation: The disturbance of the orbit of one body by the gravitational pull of a larger body. In the case of comets, the giant planet Jupiter has the most perturbing influence.

Solar Wind: The stream of particles constantly flowing from the sun in all directions. The solar wind pushes some of the material from the nucleus of a comet into long, streaming tails.

Tail: The appendage composed of gas and dust that streams out from the coma under pressure from the solar wind. The tail always points away from the sun.

SUGGESTED READING

Following is a list of some books that can lead the curious reader to a greater knowledge of Halley's Comet.

BOOKS ON GENERAL ASTRONOMY

Abell, George O. *Drama of the Universe.* New York: Holt, Rinehart and Winston, 1978.

Clayton, Donald D. *The Dark Night Sky.* New York: Quadrangle Books, 1975.

Cleminshaw, Clarence H. *The Beginner's Guide to the Sky.* New York: Thomas Y. Crowell Co., 1977.

Ley, Willy. *Watchers of the Skies.* New York: Viking Press, 1963.

Mayall, R. Newton and Margaret W. Mayall, *Olcott's Field Book of the Skies.* 4th ed. New York: G.P. Putnam's Sons, 1954.

Muirden, James. *The Amateur Astronomer's Handbook.* Revised edition. New York: Thomas Y. Crowell Co., 1974.

Shipman, Harry L. *Black Holes, Quasars, and the Universe.* 2nd ed. Boston: Houghton Mifflin Co., 1980.

Sidgwick, J.B. *Amateur Astronomer's Handbook.* New York: Dover Publications, 1980.

BOOKS ABOUT COMETS

Armitage, Angus. *Edmond Halley*. London: Thomas Nelson and Sons, Ltd., 1966.

Ball, Sir Robert S. *Comets and Shooting Stars*. London: Cassel and Company Ltd., 1910.

Brandt, John C. *Comets*. San Francisco: W.H. Freeman & Co., 1981.

Brown, Peter Lancaster. *Comets, Meteorites and Men*. New York: Taplinger Publishing Co., 1974.

Chambers, George F. *The Story of Comets*. Oxford: Clarendon Press, 1909.

Ley, Willy. *Visitors From Afar*. New York: McGraw-Hill Book Company, 1969.

MacPike, Eugene Fairfield. *Hevelius, Flamsteed and Halley*. London: Taylor and Francis, Ltd., 1937.

Manuel, Frank E. *A Portrait of Isaac Newton*. Washington, DC: New Republic Books, 1974.

Moore, Patrick. *Guide to Comets*. London: Lutterworth Press, 1977.

Proctor, Mary. *The Romance of Comets*. New York: Harper and Brothers, 1926.

Richardson, Robert Shirley. *Getting Aquainted With Comets*. New York: McGraw-Hill Book Company, 1967.

Ronan, Colin A. *Edmond Halley: Genius in Eclipse*. Garden City, NY: Doubleday & Co., 1969.

Seargent, David A. *Comets: Vagabonds of Space*. Garden City, NY: Doubleday & Co., 1982.

Westfall, Richard S. *A Biography of Isaac Newton*. London: Cambridge University Press, 1980.

Yeomans, Donald K. *The Comet Halley Handbook: An Observer's Guide*. Washington, DC: International Halley's Watch, 1981.

ADVANCED READING

Aristotle. *Meteorology*. In Great Books of the Western World, 8:445. Chicago: Encyclopædia Britannica, 1952.

Baxter, J., and T. Atkins. *The Fire Came By*. Garden City, NY: Doubleday & Co., 1976.

Delsemme, A.H., ed. *Comets, Asteroids, Meteorites: Interrelations, Evolution and Origins*. Toledo, OH: University of Toledo Press, 1977.

Drake, S., and C.D. O'Malley. *The Controversy on the Comet of 1618*. Philadelphia: University of Pennsylvania Press, 1960.

Dreyer, J.L.E. *A History of Astronomy From Thales to Kepler*. New York: Dover Press, 1953.

Goodavage, J.F. *The Comet Kohoutek*. New York: Pinnacle Books, 1973.

Hellman, C.D. *The Comet of 1577: Its Place in the History of Astronomy*. New York: AMS Press, 1944.

Herschel, J.F.W. *Outlines of Astronomy*. London: Longmans and Green, 1871.

Hoyle, F., and N.C. Wickramsinghe. *Lifecloud: The Origin of Life in the Universe*. New York: Harper and Row, 1978.

Lyttleton, R.A. *The Comets and Their Origin*. Cambridge University Press, 1953.

Marsden, B.G. "Comet Halley and History." In Neugebauer, M., Yeomans, D.K., Brandt, J.C., and Hobbs, R.W., *Space Missions to Comets*. Washington, DC: NASA, 1977.

Newton, Isaac. *Principia*. Berkeley: University of California Press, 1962.

Olivier, C.P. *Comets*. Baltimore: Williams and Wilkins, 1930.

Proctor, M., and A.C.D. Crommelin. *Comets: Their Nature, Origin and Place in the Science of Astronomy*. London: Technical Press, 1937.

Richter, N.B. *The Nature of Comets*. New York: Dover Press, 1963.